13 in 1
NO GRID SURVIVAL PROJECTS BIBLE

2500+Days of Revolutionary DIY Prepper Strategies to Achieve Energy, Build Shelters, Secure Food, and Ensure Security in Any Disaster!

IRON ROOT OUTDOOR

Copyright © 2024 by IRON ROOT OUTDOOR

All rights reserved. No part of this book may be reproduced, distributed, or transmitted in any form or by any means, including photocopying, recording, or other electronic or mechanical methods, without the prior written permission of the publisher, except in the case of brief quotations embodied in critical reviews and certain other noncommercial uses permitted by copyright law. For permission requests, write to the publisher, addressed "Attention: Permissions Coordinator," at the address below.

Table of Contents

IRON ROOT OUTDOORS: BRAND BIOGRAPHY 6

Introduction 7

Book 1: Fundamentals of Off-Grid Survival 9

I. MENTAL AND PHYSICAL PREPAREDNESS 9
II. INTEGRATION OF MENTAL AND PHYSICAL PREPAREDNESS 11
III. ESSENTIAL TOOLS AND RESOURCES 12
IV. LONG-TERM OFF-GRID SUSTAINABILITY 13
V. OFF-GRID WASTE MANAGEMENT 15

BOOK 2: Comprehensive DIY Survival Projects 20

I. OVERVIEW 20
II. THE IMPORTANCE OF A SELF-SUFFICIENT LIFESTYLE 21
III. PLANNING AND ORGANIZING FOR SELF-SUFFICIENCY 22
IV. ORGANIZING FOR SUSTAINABLE LIVING 23
V. FIRE CRAFTING MANAGEMENT 24
VI. WILDERNESS FISHING AND TRAPPING 26

BOOK 3: Alternative Energy Systems 28

I. SOLAR ENERGY 28
II. WIND ENERGY 31
III. HYDRO POWER 32

BOOK 4: Constructing Durable Shelters 37

I. CONTEMPORARY CONSTRUCTION SUPPLIES 37
II. INNOVATIVE CONSTRUCTION MATERIALS 38
III. BUILDING METHODS 38
IV. EMERGENCY SHELTER CONSTRUCTION 40

BOOK 5: Self-Sufficient Food Production and Preservation 46

I. SUSTAINABLE GARDENING TECHNIQUES 46
II. ORGANIC FARMING PRACTICES 47
III. PERMACULTURE DESIGN PRINCIPLES 49

IV. Water Conservation in Gardening 51
V. Educating and Engaging the Community: 51
VI. Food Preservation 52
VII. Root Cellars 57
VIII. Foraging Edible Plants 58
IX. Water Bath Canning and Pressure Canning 64

BOOK 6: Water Purification and Resource Management 92

I. Basic Water Purification Techniques 93
II. Mechanical Filtration 93
III. Chemical treatment 94
IV. UV and Solar Purification 96
V. Collecting and Storing Water 97

BOOK 7: Health and Wellness iN a No-Grid World 100

I. First Aid 100
II. Maintaining Physical and Mental Health 102
III. Natural Remedies and Medicinal Plants 105

BOOK 8: Security and Self-Defense Strategies 107

I. Home Security Overview 107
II. Home Security Alarm Systems 108
III. Self-Defense Basics 111
IV. Tools of Self Defense, Protection and Other Tools 113

BOOK 9: Navigating and Communicating in Off-Grid Scenarios 116

I. Land Navigation Techniques 116
II. Off-Grid Communication 118
III. Emergency Communications 120

BOOK 10: Managing Economic and Social Changes 123

I. Introduction to Bartering 123
II. Introduction to Trading 124

BOOK 11: Emergency Preparedness Checklist 126

Book 12: Navigating the Uncharted Territories of Off-Grid Living...131

BOOK 13: Printable Survival Guides for Specific Scenarios 133

Conclusion 136

IRON ROOT OUTDOORS: BRAND BIOGRAPHY

Iron Root Outdoors was founded by a group of survivalists, environmentalists, and outdoor enthusiasts who embarked on an extraordinary adventure. During an extreme expedition in one of the planet's most rugged and inhospitable regions, they encountered unforeseen challenges that pushed their endurance, ingenuity, and survival skills to the limit. This transformative experience led them to a shared vision: to establish a brand that encapsulates the strength, resilience, and deep connection with nature they had encountered.

Rooted in the philosophy of living off-grid and away from modern civilization, Iron Root Outdoors is committed to empowering adventurers, preppers, and nature lovers. The brand represents solidity and stability, qualities symbolized by the ironclad determination of our community and the deeply rooted wisdom of nature itself.

The firsthand experiences of our founders have influenced every item in our catalog, from durable survival gear to eco-friendly outdoor essentials. These products are crafted to the highest standards of quality, functionality, and ecological awareness, reflecting our dedication to environmental stewardship and sustainability.

INTRODUCTION

Living off-grid represents a profound shift away from conventional lifestyles. It involves disconnecting from mainstream utilities like electricity, water, and gas and relying instead on renewable resources and sustainable practices. Beyond its practical implications, off-grid living embodies a philosophy that resonates with individuals seeking autonomy, self-sufficiency, and harmony with nature.

At its core, the philosophy of living off-grid embraces simplicity and minimalism. It prioritizes reducing one's ecological footprint by consuming less and embracing alternative technologies such as solar panels, wind turbines, and rainwater harvesting systems. This lifestyle minimizes reliance on external resources, fostering a deeper connection to the environment and encouraging responsible stewardship of natural resources. Moreover, off-grid living embodies a spirit of independence and self-reliance. It empowers individuals to take control of their lives, liberating them from dependence on centralized infrastructures and societal norms. This autonomy extends beyond physical necessities, encompassing a sense of psychological freedom that allows individuals to align their lifestyle choices with their values and aspirations.

The philosophy of off-grid living also emphasizes community and collaboration. While autonomy is central, off-grid communities often thrive on shared knowledge, resources, and mutual support. By pooling resources and skills, individuals can create resilient and sustainable communities that prioritize collective well-being and resilience in the face of challenges. Off-grid living encourages a holistic approach to life, fostering mindfulness and a connection with the natural world. Freed from the distractions of modern life, individuals have the opportunity to cultivate meaningful relationships, pursue creative endeavors, and engage in activities that nourish the body, mind, and spirit.

In summary, the philosophy of living off-grid encompasses values of simplicity, autonomy, community, and mindfulness. It offers a pathway to a more sustainable and fulfilling way of life, rooted in harmony with nature and self-reliance. While off-grid living may not be suitable for everyone, its principles hold valuable lessons for individuals seeking a deeper connection to themselves, their communities, and the world around them. Within these pages, you'll find everything you need to embark on and thrive in your journey away from traditional grids. Starting with Book 1, you'll lay the groundwork by understanding the philosophy behind living off-grid, fortifying your mental and physical preparedness, and gathering essential tools and resources.

As you progress to Book 2, you'll delve into hands-on survival projects, guiding you through comprehensive DIY endeavors and empowering you to plan for self-sufficiency. Book 3 shifts the focus to harnessing renewable energy, whether it's solar, wind, or hydro power, ensuring you can sustainably power your off-grid lifestyle. In Book 4, you'll learn to construct resilient shelters, using the best materials and techniques to create a safe haven in any environment. The journey continues in Book 5, where the focus shifts to food, teaching you sustainable gardening techniques and food preservation methods to keep your pantry stocked year-round.

Water, a vital resource, takes center stage in Book 6, where you'll master purification, filtration, and efficient collection and storage techniques. Book 7 prioritizes health and wellness, covering everything from first aid to mental resilience in a grid-free world.

Security becomes paramount in Book 8, where you'll learn to safeguard your home and yourself in any situation. Book 9 equips you with essential navigation and communication skills for off-grid scenarios, ensuring you stay connected and informed.

Managing economic and social changes takes precedence in Book 10, emphasizing the importance of collaboration and community building. Book 11 provides a handy checklist for emergency preparedness, while Book 12 ventures into the uncharted territories of off-grid living, offering insights and guidance for navigating the unique challenges and opportunities this lifestyle presents.

Finally Book 13 offers printable survival guides for quick reference. With this guide, you're not just learning to survive off-grid – you're thriving. So, dive in with confidence and embrace the freedom of living off the grid!

BOOK 1:
FUNDAMENTALS OF OFF-GRID SURVIVAL

I. Mental and Physical Preparedness

Living off-grid demands more than just embracing alternative technologies; it requires a robust mindset and physical readiness to navigate challenges and uncertainties. Whether you're facing a power outage or a natural disaster, mental and physical preparedness are key to thriving in off-grid environments.

Mental Preparedness

Off-grid living often means encountering unexpected obstacles. Cultivating resilience equips individuals to adapt and bounce back from setbacks. Practice positive thinking, problem-solving skills, and stress management techniques to build mental toughness.

Stay present and attuned to your surroundings. Mindfulness enhances situational awareness, allowing you to identify potential risks and opportunities for improvement. Regular mindfulness practices such as meditation or deep breathing exercises can sharpen your focus and intuition.

Off-grid living may entail isolation and limited social interactions. Developing emotional intelligence fosters healthier relationships and effective communication within off-grid communities. Practice empathy, active listening, and conflict resolution skills to nurture meaningful connections.

Physical Preparedness

Fitness and Endurance: Off-grid living often entails physical labor and outdoor activities that demand a high level of fitness and endurance. From chopping wood to tending gardens, individuals must prioritize regular exercise and strength training to build a strong, resilient body capable of meeting the demands of off-grid life. Regular exercise not only improves cardiovascular health but also enhances muscular strength and endurance, essential for tasks such as carrying heavy loads and performing manual labor. Strength training, including exercises such as lifting weights and bodyweight exercises, helps build muscle mass and increase overall strength, making everyday tasks easier and more manageable. Furthermore, endurance training, such as running, hiking, or cycling, improves stamina and resilience, enabling individuals to work for extended periods without fatigue. By prioritizing fitness and endurance, individuals can optimize their physical capabilities and thrive in the demanding environment of off-grid living.

Survival Skills: In addition to physical fitness, off-grid living requires proficiency in essential survival skills that can be lifesaving in emergency situations. First aid, fire-building, and navigation are just a few examples of critical skills that individuals should master to ensure their safety and well-being. First aid training equips individuals with the knowledge and skills to provide immediate medical assistance in the event of injury or illness. From treating minor cuts and burns to administering CPR and managing more serious injuries, first aid skills are invaluable in mitigating the impact of emergencies and saving lives. Fire-building skills are essential for cooking, heating, and signaling in off-grid environments. Individuals should learn various fire-starting techniques, such as using matches, lighters, flint and steel, and friction methods, and practice building and maintaining fires under different conditions. Navigation skills are crucial for finding one's way in unfamiliar terrain and avoiding getting lost. Whether using a map and compass or relying on natural landmarks and celestial navigation, individuals should learn how to navigate effectively and confidently in off-grid environments. Investing in comprehensive training courses and regularly practicing hands-on skills are essential for maintaining proficiency in survival skills. By honing these essential abilities, individuals can enhance their readiness and resilience in off-grid living.

Nutrition and Self-Sufficiency:

Off-grid living often requires self-sufficiency in food production, making nutrition a critical aspect of physical preparedness. Sustainable gardening, permaculture, and food preservation techniques are essential skills for cultivating a reliable food source and ensuring adequate nutrition in off-grid environments.

Sustainable gardening involves growing fruits, vegetables, herbs, and other crops using organic and environmentally friendly methods. Permaculture principles, such as companion planting, polyculture, and soil conservation, help maximize yield while minimizing environmental impact. Additionally, individuals should learn food preservation techniques such as canning, drying, and fermentation to store excess produce for future use. These methods help extend the shelf life of perishable foods and reduce waste, ensuring a steady food supply throughout the year.

Prioritizing nutrient-dense foods is essential for supporting overall health and vitality in off-grid environments. Individuals should focus on incorporating a variety of fruits, vegetables, whole grains, lean proteins, and healthy fats into their diet to meet their nutritional needs and maintain optimal physical and mental performance.

II. Integration of Mental and Physical Preparedness

Holistic Health

Mental and physical well-being are intricately connected, forming the foundation of resilience in off-grid living. Adopting a holistic approach to health involves nurturing both mind and body, recognizing their interdependence in sustaining overall vitality.

Achieving Holistic Health

Nature offers a sanctuary for rejuvenation and reflection. Spend time outdoors engaging in activities like hiking, gardening, or simply immersing yourself in natural surroundings. Connecting with nature fosters a sense of peace and connection with the environment. Practice relaxation techniques such as meditation, deep breathing exercises, or yoga to promote mental clarity and emotional balance.

Nutrition plays a pivotal role in supporting physical health and mental acuity. Opt for nutrient-dense foods such as fruits, vegetables, whole grains, and lean proteins. Cultivate a diverse and sustainable diet to fuel your body and mind for the demands of off-grid living.

Continuous Learning and Adaptation

Off-grid living is a dynamic journey characterized by constant learning and adaptation. Embracing this mindset of growth and innovation enables individuals to navigate challenges effectively and thrive in ever-changing environments.

Approach off-grid living with a sense of curiosity and openness to new experiences. Seek out opportunities for learning, whether through books, workshops, or hands-on experiences. Embrace diverse perspectives and alternative approaches to problem-solving.

Expand your repertoire of skills and knowledge relevant to off-grid living. Explore topics such as renewable energy systems, permaculture techniques, wilderness survival, and emergency preparedness. Investing in comprehensive training courses equips you with the tools and expertise needed to handle diverse situations confidently.

View challenges as opportunities for growth rather than obstacles to overcome. Embrace discomfort and uncertainty as catalysts for personal and collective development. Cultivate resilience by adapting to changing circumstances with creativity and resourcefulness.

Mental and physical preparedness are essential pillars of off-grid survival. By cultivating resilience, mindfulness, and practical skills, individuals can thrive in off-grid environments, embracing the challenges and rewards of self-sufficiency and sustainability. Whether you're a seasoned off-grid enthusiast or embarking on this lifestyle for the first time, investing in your mental and physical well-being is key to success and fulfillment in off-grid living.

III. Essential Tools and Resources

Transitioning to off-grid living requires careful consideration of the tools and resources necessary to thrive independently. In this guide, we explore the fundamental tools and resources essential for success in off-grid environments, simplifying complex concepts for easy understanding.

Harnessing solar energy is a cornerstone of off-grid living, providing a sustainable and renewable source of power. Solar panels, inverters, and batteries comprise a basic solar power system, allowing individuals to generate and store electricity for various needs. Invest in high-quality solar equipment tailored to your energy requirements and geographical location to maximize efficiency and reliability.

Access to clean water is essential for off-grid living, particularly in remote or arid locations. Implementing water harvesting and filtration systems enables individuals to collect, purify, and store rainwater or groundwater for domestic use. Utilize techniques such as rainwater harvesting, gravity-fed filtration, and UV sterilization to ensure a safe and reliable water supply, reducing dependence on external sources.

Food sustainably is a core aspect of off-grid self-sufficiency, reducing reliance on grocery stores and commercial agriculture. Acquire essential gardening tools such as shovels, hoes, and watering cans to cultivate and maintain productive gardens. Incorporate permaculture principles and companion planting techniques to optimize space and maximize yield, fostering resilience and food security in off-grid environments.

Maintaining communication networks is crucial for safety and connectivity in off-grid settings, especially in emergencies or remote locations. Invest in reliable communication devices such as two-way radios, satellite phones, or handheld GPS units to facilitate communication and navigation. Prioritize devices with long battery life, rugged construction, and compatibility with off-grid infrastructure to ensure seamless connectivity in challenging environments.

Off-grid living necessitates preparedness for unexpected emergencies or natural disasters. Build a comprehensive emergency kit containing essential supplies such as first aid supplies, non-perishable food items, water purification tablets, and emergency shelter materials. Tailor your emergency kit to address specific risks and hazards prevalent in your off-grid location, ensuring readiness to handle unforeseen challenges effectively.

Managing waste responsibly is vital for preserving environmental integrity and minimizing ecological impact in off-grid settings. Implement sustainable waste management solutions such as composting toilets, recycling systems, and vermiculture bins to minimize waste generation and promote resource conservation. Adopt eco-friendly practices to reduce, reuse, and recycle materials, fostering a culture of environmental stewardship and sustainability.

Cooking is a fundamental aspect of daily life, even in off-grid settings. Invest in off-grid cooking equipment such as portable stoves, outdoor grills, and solar ovens to prepare meals efficiently and sustainably. Consider alternative fuel sources such as propane, wood, or biomass pellets for cooking, minimizing reliance on electricity or gas. Explore innovative cooking methods like Dutch oven cooking or solar cooking to conserve energy and embrace self-sufficiency in off-grid kitchens.

Creating comfortable and resilient shelter is essential for off-grid living, providing protection from the elements and ensuring a safe and secure living environment. Consider alternative housing options such as tiny houses, yurts, or earthbag dwellings that are well-suited to off-grid lifestyles. Utilize sustainable building materials like recycled wood, straw bales, or adobe bricks to construct durable and eco-friendly shelters. Incorporate passive solar design principles to optimize energy efficiency and thermal comfort in off-grid homes, reducing reliance on heating and cooling systems.

Transportation is a critical consideration for off-grid living, facilitating mobility and access to essential resources and services. Evaluate off-grid transportation solutions such as bicycles, electric bikes, or off-road vehicles suited to rugged terrain. Prioritize fuel-efficient and low-impact vehicles powered by alternative fuels like biodiesel or electricity to minimize environmental footprint. Explore alternative modes of transportation such as walking or horseback riding for short distances, promoting physical activity and reducing reliance on fossil fuels.

Managing finances effectively is essential for sustainable off-grid living, ensuring stability and resilience in the face of economic uncertainties. Develop a comprehensive financial plan that accounts for off-grid expenses such as equipment purchases, infrastructure maintenance, and renewable energy systems. Implement budgeting strategies to track income and expenses accurately, prioritizing investments in essential tools and resources for off-grid self-sufficiency. Explore alternative income streams such as freelancing, homesteading, or remote work to generate revenue while living off-grid, fostering financial independence and flexibility.

Continuous learning and skill development are essential for success in off-grid living, empowering individuals to adapt to evolving challenges and opportunities. Invest in off-grid education and training programs that cover a range of topics relevant to self-sufficiency and sustainability. Learn practical skills such as carpentry, wilderness survival, renewable energy systems, and natural building techniques through hands-on workshops or online courses. Embrace lifelong learning as a core value, cultivating a diverse skill set that enhances resilience and innovation in off-grid environments.

Building strong and resilient communities is integral to thriving in off-grid living, fostering collaboration, support, and mutual aid among like-minded individuals. Engage with local off-grid communities or form your own community of off-grid enthusiasts to share knowledge, resources, and experiences. Participate in community events, skill-sharing workshops, and cooperative projects that promote solidarity and cooperation in off-grid living. Cultivate strong interpersonal relationships based on trust, respect, and shared values, enriching the social fabric of off-grid communities and enhancing collective well-being.

Off-grid living encompasses a diverse range of considerations, from essential tools and resources to financial management, education, and community building. By embracing holistic approaches to off-grid living, individuals can cultivate resilience, sustainability, and harmony with nature in their pursuit of self-sufficiency and independence. Whether you're a seasoned off-grid enthusiast or embarking on this lifestyle for the first time, prioritizing these fundamental aspects of off-grid living will empower you to thrive in diverse and dynamic off-grid environments.

IV. Long-Term Off-Grid Sustainability

Living off-grid offers a unique opportunity to disconnect from conventional utilities and embrace a lifestyle rooted in self-sufficiency and sustainability. However, achieving long-term sustainability in off-grid living requires careful planning, dedication, and a commitment to practices that support both the environment and community resilience.

Embracing Permaculture Principles:

Permaculture, derived from "permanent agriculture," offers a holistic approach to designing sustainable systems that mimic natural ecosystems. Key principles include:

1. Observation and Interaction: Understanding local ecosystems and patterns to design resilient systems.
2. Catch and Store Energy: Utilizing renewable energy sources such as solar, wind, and hydroelectric power.
3. Apply Self-Regulation and Accept Feedback: Adapting to feedback from the environment and continuously improving systems.
4. Use and Value Renewable Resources and Services: Prioritizing renewable resources and reducing reliance on finite ones.
5. Design from Patterns to Details: Designing systems that work harmoniously with natural patterns and cycles.
6. Integrate Rather than Segregate: Creating mutually beneficial relationships between different elements in the system.
7. Use Small and Slow Solutions: Implementing changes gradually to ensure long-term effectiveness.
8. Use and Value Diversity: Encouraging biodiversity to enhance resilience and productivity.
9. Use Edges and Value the Marginal: Maximizing the productivity of edges and marginal spaces in the ecosystem.
10. Creatively Use and Respond to Change: Embracing change as an opportunity for innovation and adaptation.

Regenerative Agriculture Techniques:

Regenerative agriculture focuses on restoring soil health, increasing biodiversity, and enhancing ecosystem resilience. Key techniques include:

1. No-Till Farming: Minimizing soil disturbance to preserve soil structure and microbial diversity.
2. Cover Cropping: Planting cover crops to protect soil from erosion, suppress weeds, and improve soil fertility.
3. Crop Rotation: Rotating crops to prevent soil depletion and nutrient imbalances.
4. Composting: Recycling organic waste to create nutrient-rich compost for soil amendment.
5. Agroforestry: Integrating trees and shrubs into agricultural systems to enhance biodiversity and provide additional ecosystem services.
6. Holistic Grazing Management: Rotating livestock to mimic natural grazing patterns and improve soil health.
7. Water Harvesting: Collecting and storing rainwater for irrigation and other purposes.

Community Resilience-Building Initiatives:

Building resilience within off-grid communities involves fostering strong social networks, sharing resources, and developing collaborative solutions. Key initiatives include:

1. Community Gardens: Establishing communal gardens to promote food security and encourage collaboration among residents.

2. Skill Sharing Workshops: Organizing workshops to exchange knowledge and skills related to sustainable living practices.
3. Barter and Trade Networks: Facilitating exchanges of goods and services within the community to reduce reliance on external markets.
4. Emergency Preparedness Plans: Developing plans to address potential emergencies and disasters, such as natural disasters or supply chain disruptions.
5. Alternative Economic Systems: Exploring alternative economic models, such as local currencies or time banking, to strengthen local economies and reduce dependence on global financial systems.
6. Community Renewable Energy Projects: Collaborating on renewable energy projects to increase energy independence and resilience.

Achieving long-term sustainability and self-sufficiency in off-grid living requires a multifaceted approach that incorporates permaculture principles, regenerative agriculture techniques, and community resilience-building initiatives. By embracing these strategies and fostering a spirit of collaboration and innovation, off-grid communities can thrive in harmony with the environment while enhancing their resilience to external challenges.

V. Off-Grid Waste Management

In wilderness settings, proper waste management is essential for preserving the environment and ensuring the health and safety of both humans and wildlife. Off-grid waste management involves strategies to responsibly handle waste and sanitation without access to traditional infrastructure. This guide outlines key steps and practices for managing waste off-grid, including composting human waste, recycling materials, and minimizing environmental impact.

In this guide, we'll explore effective strategies for managing waste and sanitation off-grid, including composting human waste, recycling materials, and minimizing environmental impact.

Understanding Off-Grid Waste Management

Understanding Off-Grid Waste Management is pivotal for anyone venturing into remote areas where traditional infrastructure is absent. Here's a breakdown of key concepts:

- Environmental Impact Awareness: Recognize that every action, from waste disposal to personal hygiene, can affect the delicate balance of nature. Understanding the consequences of improper waste management is the first step towards responsible behavior.
- Resourcefulness: Off-grid waste management demands resourcefulness. It involves making the most of limited resources, repurposing materials, and adopting sustainable practices to minimize waste generation.
- Adherence to Leave No Trace Principles: Familiarize yourself with the Leave No Trace principles, which provide guidelines for minimizing environmental impact in outdoor settings. These principles include disposing of waste properly, leaving natural objects undisturbed, and respecting wildlife.
- Legal and Regulatory Considerations: Be aware of local regulations and guidelines governing waste management in wilderness areas. Different regions may have specific rules regarding

waste disposal, sanitation, and protection of natural resources. Compliance with these regulations is essential for preserving the integrity of the environment.
- Community Engagement and Education: Off-grid waste management is not solely an individual responsibility but a collective endeavor. Engage with fellow travelers, share knowledge about sustainable practices, and encourage responsible behavior to foster a culture of environmental stewardship.
- Continuous Learning and Adaptation: The dynamics of off-grid waste management are ever-evolving. Stay informed about new technologies, innovative solutions, and best practices in waste reduction and sanitation. Adapt your approach based on feedback, lessons learned, and advancements in the field.

By understanding these fundamental aspects of off-grid waste management, individuals can contribute to the preservation of wilderness areas while enjoying the beauty and solitude of remote landscapes.

Composting Human Waste

- Composting human waste is an integral aspect of off-grid waste management, particularly in wilderness settings where traditional sewage systems are unavailable. Here's a comprehensive overview of composting human waste:
- Hygiene and Safety: Prioritize hygiene and safety when composting human waste. Use designated composting toilets or dig cat holes at least 200 feet away from water sources, trails, and camping areas to prevent contamination.
- Materials Needed: Gather necessary materials, including a composting toilet or digging tools, organic cover materials such as sawdust or peat moss, and a sturdy container or compost pile for the composting process.
- Proper Collection: Ensure proper collection of human waste by using the designated toilet facilities or digging cat holes according to Leave No Trace principles. Educate all members of your group on the importance of proper waste disposal and hygiene practices.
- Covering and Mixing: After depositing human waste, cover it with organic materials such as sawdust or peat moss to facilitate decomposition and minimize odor. Mixing the waste with organic cover materials also helps balance carbon and nitrogen ratios for optimal composting.
- Aeration and Moisture Control: Maintain proper aeration and moisture levels within the compost pile or toilet system to support microbial activity and accelerate decomposition. Turn or stir the compost regularly to promote airflow and prevent anaerobic conditions.
- Temperature and Time: Monitor the temperature of the compost pile, as it should reach temperatures between 120°F and 140°F to effectively kill pathogens and ensure safe composting. Allow sufficient time for the composting process to complete, typically several months to a year, depending on environmental conditions and composting methods used.
- Final Use: Once the composting process is complete and the compost has fully decomposed, it can be safely used as fertilizer for non-edible plants in wilderness areas. Avoid using compost derived from human waste on edible crops to minimize health risks.
- Maintenance and Monitoring: Regularly inspect and maintain composting toilet systems, ensuring they remain functional and sanitary. Monitor compost piles for proper decomposition and address any issues such as odor or excessive moisture promptly.

Composting human waste is a sustainable and environmentally friendly solution for managing waste in wilderness settings. By following proper techniques and guidelines, individuals can contribute to minimizing environmental impact while promoting hygiene and sanitation in remote areas.

Recycling Materials

Recycling materials in wilderness settings is essential for reducing waste and minimizing environmental impact. Here's a guide on how to effectively recycle materials while off-grid:

- Preparation and Planning: Before your wilderness trip, prioritize items that are reusable or recyclable. Choose durable, multi-purpose products with minimal packaging to minimize waste generation from the outset.
- Sorting and Separation: Implement a system for sorting and separating recyclable materials from general waste during your trip. Use designated containers or bags to collect materials such as plastic bottles, aluminum cans, glass jars, and metal objects.
- Compact and Pack Out: Compact recyclable materials to save space and weight in your pack. Flatten plastic bottles, crush aluminum cans, and stack glass jars to maximize efficiency. Pack out recyclables with you to dispose of them properly in recycling facilities once you return from your trip.
- Educate and Encourage: Educate fellow travelers about the importance of recycling and the impact of waste on the environment. Encourage everyone to participate in recycling efforts and take responsibility for their waste by sorting materials correctly and packing out recyclables.
- Creative Reuse and Repurposing: Explore creative ways to reuse and repurpose items during your trip to minimize waste. Repurpose plastic containers for storage, use glass jars as drinking glasses or food containers, and turn old clothing into cleaning rags or insulation.
- Leave No Trace: Adhere to Leave No Trace principles when recycling materials in wilderness settings. Minimize disturbance to natural areas, avoid leaving behind any waste or recycling, and leave the environment cleaner than you found it.
- Community Collaboration: Collaborate with local organizations, parks, or wilderness agencies to establish recycling programs or initiatives in remote areas. Work together to raise awareness, provide resources, and facilitate proper waste management practices among visitors and residents alike.
- Continuous Improvement: Continuously evaluate and improve your recycling practices based on feedback, observations, and lessons learned. Explore innovative solutions, technologies, and partnerships to enhance recycling efforts and minimize environmental impact in wilderness settings.

In the realm of off-grid survival, minimizing environmental impact isn't just a choice—it's a necessity. By integrating sustainable practices into our wilderness endeavors, we not only preserve the natural beauty of remote landscapes but also ensure our own long-term viability in these environments. Through conscientious actions, from waste management to responsible recreation, we can safeguard the delicate balance of nature and leave behind a legacy of respect and stewardship for generations to come..

Advanced Techniques and Technologies

By leveraging advanced techniques and technologies for off-grid waste management, individuals and communities can mitigate environmental impact, promote resource efficiency, and enhance resilience in remote environments.

- Portable Incinerators: Utilize portable incinerators designed for off-grid use to safely burn combustible waste materials, reducing volume and minimizing environmental impact. These compact devices can efficiently incinerate organic waste, paper products, and other burnable materials, producing ash residue that can be safely disposed of or used as fertilizer.
- Bio-Digesters: Implement bio-digesters to convert organic waste into biogas and nutrient-rich compost in off-grid settings. These anaerobic digestion systems break down organic matter through microbial action, producing methane gas for energy generation and digestate for soil enrichment. Bio-digesters offer a sustainable solution for managing organic waste while generating renewable energy and improving soil fertility.
- Composting Toilets with Solar Ventilation: Install composting toilets equipped with solar-powered ventilation systems to facilitate odor control and accelerate decomposition of human waste in off-grid environments. These self-contained units use natural processes to break down solid waste into compost, reducing reliance on water-based sanitation systems and minimizing environmental impact.
- Greywater Filtration Systems: Implement greywater filtration systems to treat and recycle wastewater from sinks, showers, and laundry facilities in off-grid settings. These advanced filtration systems remove contaminants and pathogens from greywater, allowing for safe reuse in irrigation, landscaping, or non-potable applications. Greywater recycling reduces water consumption and minimizes strain on local water sources in remote areas.
- Portable Waste-to-Energy Converters: Utilize portable waste-to-energy converters to convert non-recyclable waste materials into heat or electricity for off-grid power generation. These innovative devices employ thermal or mechanical processes to convert waste into usable energy, reducing landfill volume and offsetting reliance on traditional energy sources.
- Smart Waste Monitoring Systems: Deploy smart waste monitoring systems equipped with sensors and data analytics capabilities to track waste generation, optimize collection routes, and minimize resource consumption in off-grid environments. These digital solutions provide real-time insights into waste management operations, enabling proactive decision-making and resource allocation to enhance efficiency and sustainability.
- Biomass Gasification: Explore biomass gasification technologies to convert organic waste materials, such as agricultural residues or woody biomass, into syngas for heating, cooking, or electricity generation in off-grid settings. Biomass gasifiers use thermochemical processes to break down biomass feedstock into combustible gases, offering a renewable and carbon-neutral energy source while diverting organic waste from landfills.
- Community-Based Waste Management Initiatives: Engage local communities in collaborative waste management initiatives, such as community composting programs, recycling cooperatives, or waste-to-energy projects, to address off-grid waste challenges collectively. By fostering partnerships and leveraging local resources, communities can develop sustainable waste management solutions tailored to their specific needs and priorities.

These innovative solutions offer sustainable alternatives to conventional waste management practices, contributing to a cleaner, healthier, and more sustainable future for off-grid communities and ecosystems.

BOOK 2: COMPREHENSIVE DIY SURVIVAL PROJECTS

I. Overview

In a world full of uncertainties, being able to create your own survival tools and solutions can be a literal lifesaver. Embracing a do-it-yourself (DIY) mindset empowers individuals to combine creativity with self-assurance, preparing themselves to tackle any challenge that comes their way.

Imagine finding yourself in a sudden home emergency or stranded in the wilderness, far from civilization. Knowing how to create makeshift essentials becomes incredibly valuable in such situations. Engaging in DIY rescue projects involves a wide range of activities aimed at crafting tools, equipment, and responses tailored to survival scenarios. It requires a combination of creativity, resourcefulness, and imagination to create everything from makeshift shelters to life-saving gear.

These DIY endeavors go beyond just preparing for extreme scenarios; they showcase practical skills and ingenuity. Engaging in such projects fosters a sense of empowerment and belonging for individuals, whether they live in urban areas, enjoy outdoor adventures, or simply love hands-on craftsmanship.

Let's explore some exciting DIY projects together. Imagine using everyday materials to build a solar-powered charger, a budget-friendly water filtration system, or an improvised emergency shelter.

Shelter Crafting: Basic survival skills entail the ability to fashion temporary shelters using acquired or natural resources. In emergency situations, knowing how to erect a tarp shelter, debris shelter, or makeshift tent can be crucial.

Fire-Making Techniques: Fire is a fundamental necessity for survival. Familiarize yourself with various fire-starting methods, including crafting fire starters from common items or employing traditional friction-based techniques.

Emergency Kits and Tools: Assembling compact yet comprehensive emergency kits can be lifesaving during crises. These kits may include essential items such as water purification solutions, navigation aids, and first aid supplies.

Food Sourcing and Preservation: Acquiring skills in foraging edible plants, setting traps, and preparing meals during emergencies is essential. Learn how to preserve food supplies to ensure longevity and sustenance.

Communication Devices: In a digital age where connectivity is paramount, the ability to fashion makeshift communication devices like signal devices or improvised antennas becomes indispensable when conventional means fail.

Off-Grid Energy Solutions: Harnessing renewable energy sources such as solar and wind power is key to off-grid living. Explore building your own solar panels or wind turbines to tap into sustainable energy solutions.

Embarking on these do-it-yourself projects not only equips individuals with practical survival skills but also fosters a sense of self-reliance and resilience in the face of adversity. Whether you're preparing for potential emergencies or simply indulging in hands-on creativity, these endeavors promise to enrich your life with newfound knowledge and empowerment.

II. The Importance of a Self-Sufficient Lifestyle

In life's journey, resilience shines as a symbol of self-sufficiency. Life is full of unexpected challenges and hardships, but when we have the knowledge, skills, and resources to overcome them, we can emerge stronger and more resilient. Self-sufficiency isn't just about surviving; it's about thriving in the face of adversity, equipped with the tools of self-reliance.

Self-sufficiency is a path of continuous self-improvement and growth. With each new skill acquired, each harvest reaped, and every personal challenge conquered, a sense of empowerment blossoms. The journey towards self-realization entails not just surviving, but flourishing through introspection, learning, and personal development.

Many advocates of self-sufficiency are deeply committed to environmental conservation. By growing their own food, generating their own energy, and managing resources responsibly, individuals align their lifestyle choices with sustainability principles. It's a holistic approach that harmonizes personal well-being with environmental stewardship, prioritizing the preservation of our planet for future generations.

Contrary to common misconceptions, self-sufficiency doesn't entail isolation; rather, it fosters community resilience and interconnectedness. In resilient communities, individuals come together, pooling their resources, knowledge, and support to overcome challenges collectively. It's through collaboration and mutual aid that communities thrive and endure in the face of adversity.

Choosing a life of self-reliance calls us to take back control of our lives in a world often influenced by external forces. It encourages us to assess our lifestyle choices, consumption habits, and impact on the environment. Moving towards a simpler, more meaningful existence isn't just a passing trend; it's a conscious decision to live in harmony with our values and aspirations.

The goal isn't perfection as we embark on the path of self-sufficiency. Instead, it's about making gradual progress towards a lifestyle that aligns with our values and objectives. Embracing the journey involves appreciating the small, purposeful moments that contribute to creating an independent lifestyle and enjoying the process of self-discovery along the way.

III. Planning and Organizing for Self-Sufficiency

In today's fast-paced and globally connected world, a more independent lifestyle is becoming increasingly popular. Whether driven by a desire for a more sustainable future, stability in the face of uncertainty, or more meaningful relationships in everyday life, the journey toward self-sufficiency begins with careful and gradual preparation.

Being self-sufficient is based on more than just a lifestyle; it's a thoughtful, deliberate choice to manage your resources and needs. The goal is to live in harmony with nature, encouraging self-reliance and developing the ability to deal with the inevitable challenges that life throws at us.

Embarking on the path to self-sufficiency requires careful thought and deliberate preparation. Create your own unique approach to self-sufficiency with the help of this guide:

down your goals. How do you define self-sufficiency? Is it generating your own energy, relying less on external resources, or growing your own food? Align your ideas with your aspirations. Why are you doing this? Is it to protect the environment, achieve financial independence, or live a simpler life? Identifying your values will act as a compass on your journey.

Review your current resources. Find an appropriate location, such as a balcony, backyard, or community garden. Consider factors like sunlight exposure, water access, and anything else that can affect your efforts. Evaluate your skills and knowledge. What are your strengths and where could you benefit from additional training? Continuous learning and hands-on experience are key to enrichment.

Self-sufficiency can cover a broad spectrum; prioritize areas that matter most to you to maximize your time and energy. Focus on food production, water management, renewable energy, or shelter first. Concentrating on a few priorities initially can lead to a manageable and long-term plan.

Building a self-sufficient life takes time and effort. Develop a realistic timeline that includes short-term and long-term goals. Tasks such as planting and harvesting are immediate, while long-term projects may include installing solar panels or rainwater harvesting systems.

Assess your financial situation and create a budget for your journey. While some initiatives may require upfront investment, the goal of self-sufficiency is to reduce such costs over time. Consider investments that will lead to future financial independence.

Begin with small, manageable projects to build confidence and skills. Starting with a small vegetable garden, a compost system, or a rain barrel are excellent initial steps. As you gain experience and confidence, consider expanding your projects. This approach allows for adjustments and improvements based on evolving understanding and needs.

Self-sufficiency is a journey, not a fixed destination. Embrace the unexpected, learn from challenges, and adjust your plans as necessary. Nature operates on its own schedule, and being adaptable to it is crucial for achieving self-sufficiency.

IV. Organizing for Sustainable Living

Good planning is the basis for translating planning into action, which you will need after you have a self-sustaining plan. Here are some practical suggestions to help you plan your journey:

Establish a centralized location for storing information, whether it's a digital platform or a physical binder. Be sure to include all necessary instructional materials, project plans, assessment tools, and self-reliance methods. Keep this hub updated with your progress and evolving priorities.

Setting your goals is the starting point for your journey to self-sufficiency. What does self-sufficiency mean to you? Is it learning how to grow your own food, generate your own energy, or master basic life skills? By being specific about what you want to achieve, you can lay the foundation for a strategy that aligns with your unique ambitions.

Break your journey down into manageable steps. Write down your daily, monthly, and seasonal tasks. Use calendars, whether digital or traditional, to track planting dates, harvest times, and maintenance schedules.

Allocate specific areas for particular activities based on your self-sufficiency projects. This might include creating a gardening station, a DIY workshop, or a tool shed. A well-organized workspace enhances efficiency.

Maintain meticulous records of all expenses related to your self-sufficiency efforts. This helps with budgeting and provides insights into cost-effective methods and potential areas for cost reduction.

Consistently review your strategies and practices. Reflect on what is working well, what needs adjustment, and areas where improvement or change is required. Use these evaluations to refine your self-sufficient lifestyle.

Acknowledge and celebrate milestones of progress. Celebrate accomplishments such as harvesting your first crop, completing a DIY project, or achieving a certain level of energy independence. These celebrations help to remember the positive aspects of your journey.

Achieving self-sufficiency greatly depends on your immediate environment. Take stock of your surroundings, whether it's a sprawling suburban backyard, a humble rural farmhouse, or a compact urban apartment. Adapting your self-sufficiency efforts to the space and resources at your disposal requires an understanding of your environment.

Invest in Basic Gardening Tools. One of the foundational activities for many aspiring self-sufficient individuals is growing their own food. Essential gardening tools include a hoe, rake, shovel, and pruning shears. Before purchasing seeds, compost, and soil, consider the specific needs of your chosen plants. Gardening is a skill that connects people with nature and provides them with nourishing sustenance.

The ability to build and repair common household items is a great advantage of learning carpentry. Acquire a good set of hand tools, including a hammer, screwdrivers, saw, and measuring devices. In addition to increasing your independence, mastering the basics of carpentry will give you a sense of satisfaction as you create something useful with your own hands.

To achieve self-sufficiency, it is common to produce more food than one needs at the moment. Learn the techniques of canning, drying, and fermenting, three methods of food preservation. To ensure that your harvest and homemade items last longer, invest in high-quality storage containers, such as airtight containers and glass jars.

V. Fire Crafting Management

Fire crafting and management are fundamental skills for anyone venturing into the great outdoors. Whether you're camping, hiking, or surviving in the wild, knowing how to start, build, and maintain a fire is crucial for cooking, warmth, and even signaling for help if needed. This guide will cover basic fire starting techniques, building and maintaining fires, and utilizing fire for cooking and warmth.

Basic Fire Starting Techniques:

Friction-based methods: Techniques like the bow drill or hand drill involve creating friction between two pieces of wood to generate heat and ignite tinder.

Flint and steel: Striking a piece of flint against steel produces sparks that can ignite dry tinder.

Fire starters: Carry reliable fire starters such as waterproof matches, lighters, or commercial fire starters for quick and easy ignition.

Solar methods: Utilize sunlight through magnification with a lens or by using a parabolic mirror to ignite tinder.

Building Fires:

1. Selecting the site: Choose a flat, open area away from overhanging branches, dry grass, or flammable materials. Clear the area of debris and create a fire ring if possible.
2. Fire lay: Decide on the type of fire lay depending on your needs. Common types include teepee, lean-to, and log cabin. Each has its advantages for airflow and heat distribution.
3. Gathering fuel: Collect tinder (small, dry, easily ignitable material), kindling (small sticks), and fuelwood (larger logs) in advance. Arrange them nearby in order of use.
4. Layering: Start with a base of tinder, then add kindling in increasingly larger sizes, and finally, add fuelwood on top.

Maintaining Fires:

1. Adding fuel: Regularly feed the fire with small sticks and logs to keep it burning steadily. Avoid smothering the flames by adding too much fuel at once.
2. Adjusting airflow: Control the intensity of the fire by adjusting the airflow. Adding more oxygen can make the fire burn hotter, while reducing airflow can slow it down.
3. Clearing ash: Remove ash and debris from the fire as needed to maintain airflow and prevent buildup that could smother the flames.
4. Building a sustainable fire: Continuously monitor the fire's size and adjust fuel accordingly to ensure it burns efficiently without wasting fuel.

Utilizing Fire for Cooking and Warmth:

1. Cooking over an open flame: Use a grill grate or skewers to cook food directly over the flames. Alternatively, place a pot or pan on rocks near the fire for cooking.

2. Creating coals: Let the fire burn down to hot coals, which provide an even heat source for cooking. Rake the coals into a flat surface for best results.
3. Reflectors and windbreaks: Build reflector walls or natural windbreaks to direct heat towards you and your cooking area, maximizing warmth and efficiency.
4. Safety precautions: Always practice fire safety when cooking or warming yourself near a fire. Keep a safe distance, avoid loose clothing, and never leave the fire unattended.
5. Advanced Fire Crafting Techniques:
6. Fire by friction variants: Experiment with different materials and techniques for friction-based fire starting, such as using a fire plough or a fire piston. These methods require practice and skill but can be highly effective in the right conditions.
7. Natural fire starters: Learn to identify and gather natural fire starters in the wild, such as dry grasses, pine needles, or birch bark. These materials can often be found even in damp conditions and can greatly assist in fire starting.
8. Primitive fire starting: Explore primitive methods of fire starting, such as the hand drill or the fire saw, which require only basic materials found in nature. These methods provide a deeper connection to the environment and ancient survival techniques.

Fire Management in Challenging Conditions:

1. Wet weather fires: Adapt your fire starting and management techniques for wet or damp conditions by using extra dry tinder and kindling, and constructing a platform or base to keep the fire off the wet ground.
2. Cold weather fires: In cold environments, build your fire in a pit or mound to shield it from wind and to reflect heat back towards you. Use dense, hardwoods that burn longer and hotter to combat the cold.
3. High altitude fires: Fires burn differently at high altitudes due to lower oxygen levels. Build smaller, more controlled fires and use more easily combustible materials to compensate for the reduced oxygen.

Alternative Uses of Fire:

1. Signaling: In emergency situations, fire can be used to signal for help by creating smoke during the day or using a flashlight or fire reflector at night. Learn international distress signals to communicate your need for assistance effectively.
2. Purifying water: Boiling water over a fire is an effective method for purifying it and making it safe to drink. Ensure the water reaches a rolling boil for at least one minute to kill any harmful bacteria or pathogens.
3. Creating tools: Utilize the heat of a fire to harden and shape wooden implements, such as spears or digging sticks, for hunting or gathering food.

Fire Safety and Leave No Trace Principles:

1. Extinguishing the fire: Always completely extinguish your fire before leaving the site or going to sleep. Pour water over the fire and stir the embers until they are cool to the touch. Repeat this process until there are no remaining hot spots.

2. Leave No Trace: Practice Leave No Trace principles by minimizing the impact of your fire on the environment. Use existing fire rings whenever possible or construct temporary fire pits that can be easily dismantled and naturalized after use.
3. Respecting fire bans: Be aware of any fire restrictions or bans in the area you are visiting and adhere to them strictly. In times of high fire danger, alternative cooking methods such as camp stoves may be required.

By incorporating these additional techniques and considerations into your fire crafting and management skills, you'll be better prepared to handle a variety of situations and environments in the great outdoors while minimizing your impact on the natural world.

VI. Wilderness Fishing and Trapping

In the vast expanses of wilderness, where modern conveniences are absent, mastering the art of fishing and trapping becomes essential for survival. Whether you find yourself lost in the depths of a dense forest or stranded on a deserted island, knowing how to procure food from the surrounding waters can mean the difference between sustenance and starvation. In this guide, we'll delve into the techniques and strategies for catching fish and trapping aquatic creatures in no-grid survival situations, utilizing improvised gear, strategic bait selection, and keen observational skills.

Understanding Your Environment

Before casting your line or setting traps, it's crucial to understand the aquatic ecosystem you're navigating. Take the time to observe the water bodies around you, noting any signs of fish activity such as ripples, jumping fish, or nesting areas. Understanding the behavior and habits of your target species will greatly increase your chances of success.

Improvised Fishing Gear

In a no-grid survival scenario, you won't have access to fancy fishing rods or reels. Instead, you'll need to improvise using whatever materials you have at hand. Craft a simple fishing rod using a sturdy stick as the rod, fishing line or cordage as the line, and a makeshift hook fashioned from a safety pin, paperclip, or carved wood. Alternatively, try your hand at hand-line fishing, where you'll simply attach bait to a line and cast it out into the water, using your hands to feel for bites.

Bait Selection

Choosing the right bait is crucial for attracting fish to your line or traps. In a wilderness setting, you can use a variety of natural baits such as worms, insects, grubs, or even small fish caught from the area. Experiment with different baits to see what works best for the species you're targeting, and don't be afraid to get creative with your offerings.

Identifying Optimal Fishing Locations

Finding the best fishing spots requires a keen eye and an understanding of fish behavior. Look for areas with underwater structure such as rocks, fallen trees, or submerged vegetation, as these provide hiding spots for fish and attract prey. Additionally, pay attention to water currents, as fish often congregate in areas where food is carried by the flow. By observing your surroundings and experimenting with different locations, you'll increase your chances of making a successful catch.

Trapping Aquatic Creatures

In addition to fishing, trapping can be an effective method for procuring food in wilderness settings. Construct simple traps using materials such as sticks, rocks, and natural fibers, and place them strategically in areas where you've observed signs of aquatic life. Experiment with different trap designs and bait options to see what yields the best results, and be patient as you wait for your prey to take the bait.

Mastering the art of fishing and trapping in a no-grid survival situation requires patience, resourcefulness, and a deep understanding of the natural world. By honing your observational skills, crafting improvised gear, and experimenting with different techniques, you'll be better equipped to navigate the challenges of wilderness living and ensure your continued survival in even the most remote environments.

BOOK 3:
ALTERNATIVE ENERGY SYSTEMS

I. Solar Energy

As the world increasingly moves towards sustainable living, solar energy emerges as a leading alternative energy source. By harnessing sunlight to generate electricity, solar power is both sustainable and environmentally friendly, drawing from an infinite and renewable resource pool. Let's explore the intricacies of solar energy, the foundation upon which solar systems are built.

At the core of solar energy is a simple yet ingenious concept: capturing sunlight and converting it into usable heat or electricity. This is achieved through photovoltaic cells, also known as solar panels, which are the driving force behind this technology. Inside these cells, sunlight interacts with semiconductor materials like silicon, causing the release of electrons and generating electricity through the photoelectric effect.

The concept of photovoltaic cells dates back to the discovery of the photoelectric effect in the mid-19th century. When sunlight hits the surface of a solar panel, electrons in the semiconductor material become excited, leading to the generation of electric current. This direct current (DC) is then converted into alternating current (AC) – the standard form of electricity used in homes and businesses – by an inverter.

A solar system consists of several key components, with solar panels being the primary element. These panels are made up of interconnected photovoltaic cells, and their efficiency, orientation, and placement greatly affect the system's energy production. In addition, inverters are crucial for converting DC electricity from the panels into AC power suitable for household use.

Proper installation of solar panels is essential for their effectiveness. Factors such as sunlight exposure throughout the day determine whether panels should be installed on rooftops, in attics, or on the ground. Wiring and racking systems help integrate panels into the property's electrical grid, ensuring smooth operation.

Some solar systems include battery storage to store excess energy generated during peak sunlight hours. This stored energy can then be used during periods of low sunlight or at night, ensuring a reliable and uninterrupted power supply.

Several factors affect the efficiency of solar panels, which is defined as the percentage of sunlight converted into usable energy. Panel quality, weather conditions, shading, orientation, and tilt all influence energy generation. Strategic planning and positioning are crucial to maximize efficiency and ensure optimal sunlight exposure for enhanced energy production.

Solar energy represents a sustainable and innovative solution in the search for cleaner, greener energy sources. By understanding the basics of solar power generation and harnessing technological advancements, individuals and communities can embrace solar energy as a viable pathway towards a brighter and more sustainable future.

Benefits of Solar Energy

The allure of solar energy goes beyond the fact that it is renewable. There are numerous benefits to switching to solar energy. Solar energy helps keep the planet livable by generating clean power without depleting finite resources and without contributing to global warming. By generating electricity on-site, solar energy can significantly reduce reliance on traditional power from the grid, resulting in lower electricity bills.

By installing solar panels, homes and businesses can reduce their dependence on external energy sources and the fluctuations in electricity prices. This provides them with a degree of energy autonomy.

Solar energy systems, despite the upfront costs, are investments for the long term since they provide a sustainable energy source that can increase property value.

Challenges and Future of Solar Energy

Solar energy has a bright future, but there are still challenges. Potential obstacles include high installation costs, weather-related disruptions, and limited available space. The good news is that technology is constantly improving, meaning that solar panels are becoming more affordable and more efficient.

The limitless potential of the sun becomes more apparent as we delve into the basics of solar energy. From individual solar panels on rooftops to large solar farms harvesting sunlight, the widespread adoption of solar technology and installation offers a sustainable and promising future.

Ultimately, exploring the basics of solar energy inspires a vision for a greener and more sustainable lifestyle. It invites us to harness the power of the sun not only as a source of light and warmth but as an endless source of clean energy, lighting the way to a greener tomorrow.

Solar System Design

Due to growing environmental and sustainability concerns, more and more people are turning to renewable energy sources, particularly solar power. With the right information, designing a solar system for your home is not as intimidating as it may first appear. As we delve deeper into solar system design, we simplify the process in a way that everyone can understand. By doing so, you'll have the potential to harness solar energy and contribute to a more sustainable future.

An understanding of the basics is crucial before exploring the complexities of solar power system design. Photovoltaic (PV) cells are the heart of solar power systems, which transform sunlight into usable electricity. These cells, which are typically seen in solar panels, absorb light and convert it into DC electricity. An inverter then changes the DC electricity to AC electricity, which is suitable for use in homes.

The first step in designing a solar energy system that will work is knowing how much energy you use. Conduct an in-depth home energy audit to determine your current energy consumption. Review your utility bills, identify peak usage periods, and consider energy-efficient habits to reduce overall consumption. You cannot estimate the potential output of your solar energy system without first conducting this audit.

Evaluate the solar potential of your property by considering factors such as location, roof orientation, shading, and available space. Use online tools like the Solar Energy Industries Association's Solar

Potential Calculator to estimate the amount of sunlight your property receives and its suitability for solar panels.

Based on your energy audit and solar potential assessment, determine the size of the solar system needed to meet your energy needs. Consider the types of solar panels, inverters, mounting systems, and batteries (if opting for a battery storage system) that will make up your solar system.

Choose high-quality solar panels and inverters that offer good warranties and are suitable for your climate and energy requirements. Research different brands and models to find the best fit for your budget and needs.

Consider hiring a professional solar installer to ensure proper installation and safety. Regular maintenance, such as cleaning solar panels and checking for any damage, is essential to maximize the efficiency and lifespan of your solar system.

Invest in a monitoring system to track the performance of your solar system. This will help you identify any issues early on and optimize the system for better efficiency.

Assessing Your Roof's Suitability

Before installing solar panels, it's crucial to assess whether your roof is suitable. Factors like orientation, tilt, and shading play a significant role in the system's efficiency. Roofs facing south with minimal shading are ideal, but east or west orientations can also work. If you're unsure about your roof's compatibility, consult a solar expert.

After assessing your energy needs and roof suitability, calculate the size of the solar system you require. Consider the wattage of the panels, their efficiency, and the average hours of sunlight in your area. Solar professionals and online calculators can assist in this process.

Choose the right type of solar panels for your needs, whether they're monocrystalline, polycrystalline, or thin-film. Consider factors like efficiency, aesthetics, and cost. Ensure the inverter matches your system's size and consider installing a monitoring system to track your panels' energy production.

Research solar energy incentives and regulations in your area. Many regions offer rebates, tax credits, or other financial incentives for solar panel installations. Understanding these incentives can significantly impact your system's overall cost and enhance the financial benefits of solar energy.

While solar systems require an initial investment, the long-term savings can offset the costs. Create a detailed financial plan that includes the cost of solar panels, inverters, installation, and other necessary components. Explore solar leasing programs or financing options to ease the financial burden of transitioning to solar energy.

Consider hiring a professional solar energy installation service for safety and efficiency. Look for a reputable solar energy company with experience in your area. Obtain quotes, read reviews, and check certifications to find a provider that fits your budget and needs.

Regular maintenance and optimization are essential for the performance and lifespan of your solar energy system. Clean the solar panels regularly to remove dust or debris and use the monitoring tools to keep track of the system's performance. Address any issues promptly to ensure consistent energy generation.

Adopting a solar energy system is a step toward a greener lifestyle. Make small changes in your daily routine to save energy and reduce your carbon footprint. Educate yourself and your loved ones on energy conservation and take pride in the positive impact you're making on the planet.

II. Wind Energy

Wind power represents more than just an alternative energy source; it's a symbol of independence, sustainability, and harmony with nature. As we strive for self-sufficiency, the gentle hum of wind turbines becomes the soundtrack of our daily lives, reminding us of our commitment to living in balance with the environment.

Wind power harnesses the natural forces of the wind to generate electricity. At the heart of this process are wind turbines, which convert the kinetic energy of the wind into electrical power. These turbines typically consist of a tower, blades, and a nacelle (housing the generator and other components). When the wind blows, it turns the blades, which in turn spins a shaft connected to a generator, producing electricity. Wind is a plentiful and inexhaustible resource, making it an ideal energy source for those of us committed to reducing our environmental impact and living sustainably.

By generating our own electricity from the wind, we reduce our reliance on external power sources and gain greater control over our energy supply, which is especially important in remote or isolated locations. Although there is an initial investment in setting up a wind turbine, the long-term savings on energy costs and the potential for selling excess power back to the grid make it a financially viable option for many off-grid dwellers. Once installed, wind turbines require relatively low maintenance, making them a practical choice for those of us who prefer to focus on other aspects of off-grid living.

Challenges and Considerations

While wind power offers many advantages, there are also challenges to consider. These include finding a suitable location with consistent wind speeds, navigating local zoning and permitting regulations, and managing the initial cost of installation. Wind power generation can be intermittent, so it's often used in conjunction with other energy sources or storage solutions to ensure a reliable power supply.

Preparing for Backyard Wind Power

Harnessing wind energy in your backyard can be a rewarding way to generate renewable electricity and reduce your environmental footprint. However, before you can enjoy the benefits of backyard wind power, there are several important steps you need to take to prepare:

The first step is to determine if your location has sufficient wind speed to justify a wind turbine. An average wind speed of at least 10-12 mph (4.5-5.4 m/s) is generally considered the minimum for a small wind turbine to be effective. Consider using an anemometer to measure wind speed at the proposed height of your turbine over a period of time (typically a year) to get accurate data. Select a wind turbine that is appropriate for your wind resource, energy needs, and property size. Consider whether a horizontal-axis or vertical-axis turbine is more suitable for your situation. Determine the capacity of the turbine (in kilowatts) based on your energy consumption and the average wind speed at your site. Check with your local planning and zoning department to understand any restrictions or permits required for installing a wind turbine in your area.

Be aware of any height restrictions, noise ordinances, or aesthetic concerns that may impact your ability to install a turbine. Identify the best location on your property for the turbine, considering factors such as distance from your home, proximity to obstacles (trees, buildings), and ease of access for maintenance. Plan for a sturdy foundation to support the turbine, taking into account soil type and stability. Decide whether you will connect the turbine to the grid or use it for off-grid power. Grid-connected systems may allow you to sell excess power back to the utility company.

If you opt for an off-grid system, consider the need for battery storage to ensure a consistent power supply. Plan for safety features such as lightning protection, emergency shut-off, and proper grounding. Maintenance is essential for optimal performance. Familiarize yourself with the manufacturer's maintenance recommendations and consider how you will access the turbine for repairs and upkeep. Calculate the total cost of the turbine, installation, permits, and any additional equipment (such as batteries or inverters). Look into federal, state, or local incentives, rebates, or tax credits that may help offset the cost of your wind system. Consider hiring a professional installer with experience in wind turbine systems to ensure proper setup and safety. Once installed, test the system to ensure it is operating correctly and efficiently.

III. Hydro Power

Hydro power, also known as hydropower or hydroelectric power, is a form of renewable energy that harnesses the energy of flowing or falling water to generate electricity. For individuals living off-grid, hydro power can be an invaluable resource for sustainable, self-sufficient energy production. Hydro power is based on the simple principle of converting the kinetic energy of moving water into mechanical energy, which is then converted into electrical energy through a generator. The amount of energy produced depends on the volume of water flow and the height from which the water falls, known as the head.

Types of Hydro Power Systems for Off-Grid Living

Micro hydro and Pico hydro systems represent the smaller scale of hydroelectric power generation, tailored for individual homes or small communities living off-grid. Micro hydro systems, designed to produce less than 100 kW of power, are an ideal solution for areas with a consistent flow of water, providing a sustainable and reliable source of electricity. These systems harness the kinetic energy of flowing water, converting it into electrical energy through a turbine and generator setup. The beauty of micro hydro systems lies in their ability to integrate seamlessly with the natural environment, causing minimal disruption to the surrounding ecosystem.

On an even smaller scale, Pico hydro systems are designed for very modest energy needs, typically generating less than 5 kW of power. These setups are perfect for powering a few lights or small appliances, making them a suitable choice for remote cabins, tiny homes, or any off-grid dwelling with limited energy requirements. Pico hydro systems offer a simple and cost-effective way to tap into the power of flowing water, providing a gentle nudge of electricity to those who need just a little to get by.

Both micro and Pico hydro systems embody the principles of sustainable living, offering off-grid communities a way to harness natural resources without relying on fossil fuels. Their low environmental impact and ability to provide continuous power make them an attractive option for those seeking to reduce their carbon footprint and live in harmony with nature. Whether it's the modest power of a Pico hydro system or the more substantial output of a micro hydro setup, these renewable energy solutions offer a path toward a more sustainable and self-sufficient lifestyle.

Components of a Hydro Power System

Micro hydro systems consist of several key components, each playing a crucial role in the process of converting water's kinetic energy into usable electrical power.

The journey of water in a micro hydro system begins at the intake, where water is diverted from a stream or river into a channel or pipe, setting the stage for energy generation. This water is then directed through a penstock, a pipeline that conveys the water from the intake to the turbine. The penstock is instrumental in creating the necessary pressure for energy generation, guiding the water with precision to its next destination.

At the heart of the micro hydro system lies the turbine, a marvel of engineering where the magic of energy conversion takes place. As water flow spins the turbine blades, kinetic energy is transformed into mechanical energy, a pivotal step in the generation of electricity. This mechanical energy is then transferred to the generator, which is connected to the turbine and plays a vital role in converting mechanical energy into electrical energy. The generator acts as the bridge between the mechanical and electrical realms, ensuring that the energy harnessed from the water is ready for use in powering homes and appliances.

To regulate the operation of the turbine and generator, a control system is employed, ensuring optimal performance and safety. This system monitors and adjusts the functioning of the micro hydro setup, maintaining a delicate balance between efficiency and protection. Finally, a battery bank serves as a reservoir for excess electricity generated, storing this valuable energy for use when water flow is insufficient or during peak demand periods. The battery bank provides a buffer, ensuring a steady supply of electricity even in fluctuating conditions.

Together, these components form a cohesive and effective micro hydro system, capable of providing reliable and renewable energy in off-grid settings. By tapping into the natural flow of water, micro hydro systems offer a sustainable path to energy independence, reducing reliance on traditional power sources and fostering a harmonious relationship with the environment.

Advantages

The beauty of micro hydro systems lies in their ability to tap into the natural water cycle, making them a renewable energy source that doesn't deplete the earth's resources. This renewable aspect ensures that as long as there is a consistent flow of water, the system can provide a steady and reliable supply of electricity, which is crucial for off-grid living where energy sources can often be limited or inconsistent.

One of the most appealing aspects of micro hydro systems is their low operating costs. Once the initial setup is complete, the maintenance and operational expenses are relatively minimal, making it an economically viable option for long-term energy generation. This cost-effectiveness is further enhanced by the reliability of the system, as a steady water source can ensure a continuous supply of electricity, reducing the need for frequent repairs or replacements.

Micro hydro systems are environmentally friendly, producing no direct emissions. This clean energy source aligns with the growing trend towards sustainable living, allowing individuals to reduce their carbon footprint and contribute positively to the environment. The absence of emissions also means that micro hydro systems do not contribute to air pollution, making them a healthier option for both the planet and its inhabitants.

Considerations

Water rights are a crucial aspect of setting up a micro hydro system. It's essential to ensure that you have the legal right to use the water source for energy generation. This may involve obtaining permits or licenses from local authorities or water management agencies. Ensuring that you have the proper rights in place helps avoid legal complications and ensures that the water use is compliant with local regulations.

The environmental impact of the micro hydro system must be carefully considered. The introduction of a hydroelectric system can affect local ecosystems and aquatic life, especially if the water flow is significantly altered. Conducting environmental assessments and implementing measures to minimize negative impacts are important steps in the planning process. This might include designing fish-friendly turbines or maintaining minimum flow levels in the watercourse to support aquatic habitats.

Chosen location must have a sufficient flow of water and an adequate head (the vertical drop of water) to generate energy effectively. The site should also be accessible for installation and maintenance and have a suitable area for the placement of the turbine, generator, and other components. A thorough assessment of the site's hydrological characteristics is necessary to determine its feasibility and to design a system that optimizes energy generation.

The initial investment required to set up a micro hydro system is a significant consideration. While the upfront costs can be high, including expenses for equipment, installation, and any necessary modifications to the watercourse, the long-term benefits often outweigh these costs. Micro hydro systems have relatively low operating and maintenance expenses, and they provide a renewable source of energy that can reduce reliance on fossil fuels and lower energy bills. Careful financial planning and exploring options for funding, grants, or incentives can help manage the initial investment and ensure the project's economic viability.

Installing Hydroelectric Power System

The first step in setting up a hydropower system is to assess the availability and flow rate of a nearby water source, such as a stream, river, or creek. The water source should have sufficient flow rate and drop (head) to generate adequate power for your needs. Conducting a site survey and obtaining any necessary permits or approvals from local authorities is essential before proceeding.

Once the water source is deemed suitable, the next phase is planning and designing the hydropower system. This involves determining the type of turbine best suited for the site's flow rate and head, as well as designing the intake, penstock (pipe that conveys water to the turbine), and tailrace (channel that carries water away from the turbine). Consulting with a hydro power expert or engineer can help ensure that the system is efficiently designed to meet your energy needs.

Selecting the right components is crucial for the system's success. Key components of a hydropower system include the turbine, generator, controller, and batteries for energy storage. The turbine converts the kinetic energy of flowing water into mechanical energy, which is then converted into electrical energy by the generator. The controller regulates the system's operation, and batteries store excess electricity for use when the water flow is insufficient.

Installing a hydropower system involves constructing the intake structure to divert water from the source, laying the penstock to channel water to the turbine, and setting up the turbine and generator. The installation process requires careful planning and execution to ensure that the system is safe, efficient, and environmentally friendly.

After installation, the system is connected to your off-grid home's electrical system, and it's crucial to test the system thoroughly to ensure that it operates smoothly and efficiently under various flow conditions. Adjustments may be necessary to optimize performance.

Regular maintenance is essential to keep the hydropower system running smoothly. This includes inspecting and cleaning the intake screen, checking the penstock for leaks, lubricating moving parts, and monitoring the system's performance. Proper maintenance helps extend the lifespan of the system and ensures reliable electricity generation.

Finally, it's important to consider the environmental impact of a hydro power installation. Measures should be taken to minimize disruption to the water source's ecosystem, such as ensuring fish passage and maintaining water quality.

Selecting the Right Hydro Power Turbine

Selecting the right hydro power turbine for off-grid living involves a detailed assessment of your water source and energy needs, along with careful consideration of the turbine's characteristics and environmental impact. First, evaluate the flow rate, head, and seasonal variations of your water source to determine the potential energy output and the most suitable type of turbine. The type of turbine you choose, whether it's a Pelton wheel for high-head, low-flow situations, a Francis turbine for medium-head, medium-flow scenarios, or a Kaplan turbine for low-head, high-flow conditions, will depend on the specific characteristics of your water source.

Once you've identified the right type of turbine, consider its power output in relation to your daily and seasonal energy consumption. It's important to ensure that the turbine can provide sufficient power for your off-grid home while accounting for efficiency losses in the system. Look for turbines with high efficiency ratings, as these will convert a higher percentage of water energy into electrical energy, providing more power for your needs.

Integrate smart control technology into the hydro power system. Smart controls allow for real-time monitoring and adjustment of the system's operation, optimizing efficiency and adapting to changes in water flow. This can include automatic turbine speed control, remote system monitoring, and data analysis to predict maintenance needs.

Connect the hydro power system to a battery bank or other energy storage solutions. This ensures a continuous power supply, even during periods of low water flow or maintenance. It's important to size the storage system based on your energy needs and the expected energy production from the turbine.

Wire the system to your off-grid home, ensuring that the generated electricity is safely and efficiently distributed. This may involve setting up an inverter to convert the electricity from DC to AC, installing circuit breakers, and grounding the system for safety.

The environmental impact of your hydro power system is another crucial factor. Choose a turbine that does not significantly alter the natural flow of the water or harm aquatic life. Some turbines are designed to be fish-friendly, which is an important consideration for maintaining the health of your water source.

Installation and maintenance requirements are also key considerations. Some turbines may require more technical expertise or regular maintenance, so ensure you have the skills or support to manage the system effectively. In addition, assess the upfront cost of the turbine and balance it against its expected lifespan and maintenance costs to ensure it offers good long-term value.

Lastly, check local regulations regarding water rights and environmental impact to ensure your turbine and installation comply with all legal requirements. Consulting with hydro power experts or suppliers can provide valuable insights and recommendations for selecting the most suitable turbine for your off-grid setup.

BOOK 4:
CONSTRUCTING DURABLE SHELTERS

I. Contemporary Construction Supplies

The quest for a place to call home has been a central pursuit throughout human history. From the first mud brick homes to the towering ancient stone temples, building materials have always been a way for people to express their history and way of life. This chapter explores the age-old craftsmanship of enduring residential construction, highlighting the traditional materials that have contributed to the historic splendor and lasting energy of buildings.

Clay and Loam

The adaptability and plasticity of clay and loam have influenced the creation of human settlements for centuries. These easily moldable soils blend seamlessly with the surrounding landscape, providing a pristine canvas for residential design. The sun-dried adobe bricks of ancient civilizations silently attest to the strength and durability of mud dwellings. Adobe architecture, refined in arid regions, involves mixing clay, sand, and straw to create bricks that are then sun-baked to harden. The resulting structures are highly effective in providing thermoregulatory properties, creating cool retreats in the scorching heat and maintaining warmth during winter nights. Not only is the material exquisitely tactile, but it also carries the wisdom necessary to create lasting havens embraced by the earth itself.

Stone

Stone, a symbol of strength and permanence, has been the foundation of enduring structures throughout history. From the grand fortresses of medieval Europe to the intricately carved temples of ancient India, stone has been used for more than just construction; it has been a monument to the skill of its craftsmen. The durability of stone is beyond evaluation, making it a favored material in the building of historical magnificence.

Wood

Natural materials like wood have been the go-to choice for eco-friendly construction. From ancient Japanese temples to European cottages, wooden carvings showcase the perfect harmony of human artistry and the natural beauty of wood. The inherent integrity and flexibility of wood make it an ideal material for building techniques. Wooden structures demonstrate the expertise of traditional craftsmanship, as they are constructed using wooden components that do not require nails. Beyond functionality, the inherent thermal properties of wood connect the natural aesthetics that resonate with people to the planet's environment.

Bamboo

In regions where bamboo grows abundantly, this versatile grass serves an incredibly useful architectural purpose. Bamboo's remarkable strength-to-weight ratio combined with its rapid regeneration makes it an ideal material for sustainable construction. The adaptability of bamboo has long been recognized by people in South and Southeast Asia, who have used it to build homes and bridges.

Thatch

Traditional roofing techniques, such as thatching with materials like straw, reed, or grass, have been used for a very long time. Thatched roof dwellings can be found in various cultures, from the English countryside to the arid regions of Africa. Not only is thatch aesthetically pleasing, but it also has excellent insulating properties and is long-lasting. The thatched roofs act as natural insulation, keeping the homes warm in the winter and cool in the summer. Sourced locally, these materials have a lower environmental impact. The tradition of constructing shelters and using thatch is a long-standing practice that keeps people connected and preserves their heritage.

II. Innovative Construction Materials

Recycled Plastic Bricks

Surprisingly, the sustainable housing industry is becoming a partner in the global fight against plastic pollution. Recycled plastic bricks, made from used plastic bottles and other waste materials, offer an alternative solution to overpopulation and pollution problems. These bricks are durable and strong, helping reduce the harmful amount of plastic waste that ends up in our waterways.

In this process, scrap plastic is molded into modular bricks that are easy to assemble. Recycled plastic bricks demonstrate how we can innovate to live better by turning waste into a valuable product. There is a pressing demand for affordable, eco-friendly housing, and this alternative use of plastic waste not only helps prevent environmental pollution but also provides a real solution to that problem.

Hempcrete

A new and exciting in the race for environmentally friendly building materials. Made from the woody fibers of the hemp plant, lime, and water, hempcrete is a sustainable alternative to traditional concrete. Hemp is an eco-friendly crop, as it requires minimal cultivation or pesticides to grow. The end result is a lightweight product with excellent insulating properties.

Hempcrete structures are breathable and provide a healthy and comfortable living environment. Another reason hempcrete is great for the environment is that it produces no carbon emissions. As we explore new materials, hempcrete stands out as a prime example of using plant-based solutions to construct homes in harmony with nature's sustainability cycles.

III. BUILDING METHODS

Throughout our diverse history, the constant need for shelter has often ignited the creative spirit of our fellow humans. Improvised construction techniques have allowed communities around the world to create long-lasting shelters by combining imagination and practicality in the face of limited resources or unusual constraints. In this chapter, we will explore the varied terrains of improvised building, where creativity and practicality come together to construct shelters.

Earthbag Construction

Earthbag construction stands out as a subtle and unassuming method of creating customized structures. The basic idea involves stacking hard bags filled with local materials such as clay or sand to form walls.

Wrapping the bags with barbed wire creates a strong structure that can be built in a wide range of designs.

Earthbag construction is highly regarded for its versatility, cost-effectiveness, and adaptability in different environments. This method has been used worldwide, from temporary shelters to permanent dwellings. Instead of the straight lines of a conventional house, earthbag walls can be shaped organically. Each bag, when placed in the hands of the builders, becomes a small but essential brick in constructing a sturdy home.

Tire Walls

In the quest for green homes, one innovative method involves the use of discarded tires. People have constructed eco-friendly and easy-to-assemble walls out of compacted earth and stacks of tires. Due to the abundance of these materials, recycled tires can become a building material for constructing sustainable homes over the long term.

This material is particularly well-suited for high-temperature environments due to its excellent insulating properties. The stacking of tires creates robust walls that act as a heat regulator, making the dwelling much more comfortable. By recycling waste tires into solid foundations, these structures not only provide shelter but also represent a commitment to environmentally friendly practices.

Housing in Shipping Containers

Repurposing shipping containers as dwellings creates a notable niche in the improvised building landscape, echoing the hum of global trade. These sturdy steel structures, once used to transport goods across seas, are transformed into modular housing that challenges conventional wisdom. Due to their versatility, repurposed shipping containers can be used to construct multi-story homes, offices, or even community centers.

The environmental credentials of shipping container homes are enhanced by the fact that they repurpose leftover containers, in addition to their structural soundness. Recycling these large metal structures not only reduces their negative impact on the environment but also provides an affordable option for long-lasting housing. The remarkable transformation of cargo containers into residential spaces showcases the relentless power of human creativity in the face of ever-changing challenges.

Cob Building

Cob building, an ancient construction method that originated in places like England, involves combining clay, straw, and water to create durable materials. Cob houses are known for blending in and harmonizing with their natural surroundings.

The hands-on aspect of the process makes construction a communal effort, as people come together to engage in cob building. Cob structures often feature curved walls, ornate decorations, and artistic touches, reflecting the uniqueness and creativity of their builders. When worked by skilled artisans, cob provides a means of constructing homes that are not just functional; they are works of art that honor human ingenuity and the earth.

The Natural Scaffolding for Resilient Shelters

People around the world have used bamboo as a primary building material for centuries because it is durable, resistant to conditions, and time-tested. Bamboo's versatility makes it suitable for a wide range of construction projects, from simple dwellings to majestic temples with intricate designs.

The beauty of bamboo lies in its strength and elegance. As it can be harvested sustainably, bamboo is a more environmentally friendly material than traditional building materials that offer minimal disruption. Bamboo structures represent the balance of cooperation between people and nature in the world, and the many elements involved in providing stable housing.

IV. Emergency Shelter Construction

In the wild, unforeseen circumstances can leave you exposed to the elements with no shelter in sight. In such scenarios, the ability to construct emergency shelters using natural materials becomes paramount for survival. This guide offers step-by-step instructions for building sturdy shelters from branches, leaves, and mud, along with tips for selecting safe and suitable shelter locations to maximize your chances of weathering the storm.

- Assessing Your Surroundings

Assessing your surroundings is the crucial first step in building an emergency shelter in a no-grid survival situation. Here's a detailed guide on how to do it effectively:

- Observation

Take a moment to carefully observe your surroundings. Look for features such as terrain, vegetation, water sources, and potential hazards like falling branches, steep slopes, or animal dens.

- Terrain

Choose a location that is relatively flat and free from obstacles. Avoid areas prone to flooding, as well as low-lying spots where water might accumulate during rainfall. If possible, select higher ground to minimize the risk of flooding.

- Water Source

Consider proximity to a water source. While you don't want to be too close to avoid flooding, having access to water is essential for hydration and other survival needs. Look for streams, rivers, or lakes nearby.

- Natural Shelter

Look for natural features that can provide additional shelter or protection, such as rock overhangs, dense tree canopies, or large bushes. These features can supplement your shelter or serve as a backup if needed.

- Wind Direction

Assess the prevailing wind direction in your area. Position your shelter so that the entrance is facing away from the prevailing wind to minimize drafts and maximize comfort. This will also help prevent smoke from fires blowing into your shelter.

- Sun Exposure

Consider the path of the sun throughout the day. While you want your shelter to receive sunlight during the day for warmth, you also want it to provide shade during the hottest parts of the day. Position your shelter accordingly to balance these factors.

- Visibility

Choose a location that is visible to search and rescue teams or passing aircraft, but avoid being too exposed to potential threats. A balance between visibility and concealment is ideal.

- Security

Assess the safety and security of the area. Look for signs of wildlife activity, such as tracks or droppings, and avoid areas that appear to be frequented by predators. Ensure that your chosen location is not in the path of potential hazards like falling rocks or dead trees.

- Accessibility

Consider how easily you can access your shelter location. Make sure it's within a reasonable distance from your current location, especially if you're injured or carrying heavy gear.

- Resource Availability

Take note of nearby natural resources that can be used for shelter construction, such as branches, leaves, rocks, and other materials. Having these resources readily available will make building your shelter easier and more efficient.

By carefully assessing your surroundings and considering these factors, you can choose a suitable location for building your emergency shelter and increase your chances of survival in a no-grid survival situation.

Building a Lean-To Shelter

Building a lean-to shelter is a straightforward and effective way to create protection from the elements in a no-grid survival situation. Here's a step-by-step guide on how to build one:

Materials Needed:

- Sturdy branch or pole for the main support
- Smaller branches or saplings for framing
- Large leaves, ferns, or other foliage for covering
- Additional foliage or grass for insulation (optional)

Steps:

1. Select a Location

Choose a location for your lean-to shelter that meets the criteria outlined during the assessment of your surroundings. Ensure it's close to resources like water and suitable materials for construction, but away from potential hazards.

2. Set up the Main Support

Find a sturdy branch or pole that is long enough to lean against a tree or another support structure at a 45-degree angle. Secure it in place by wedging it between rocks, tying it to a tree, or burying it in the ground.

3. Frame the Shelter

Gather smaller branches or saplings to create the frame of your lean-to. Lean them against the main support at a slight angle, spacing them out evenly to form a lattice-like structure. If necessary, use cordage or vines to lash the branches together for added stability.

4. Cover with Foliage

Collect large leaves, ferns, or other foliage to cover the frame of your lean-to shelter. Layer them over the frame, starting from the bottom and working your way up, overlapping each layer to ensure waterproofing. If large leaves are scarce, smaller foliage can be used as well.

5. Insulate the Interior (Optional)

If you have additional foliage or grass available, you can use it to insulate the interior of your shelter for added warmth and comfort. Layer the foliage on the ground inside the shelter to create a soft, insulating barrier between you and the cold ground.

6. Finishing Touches

Once your lean-to shelter is constructed and covered, take a moment to reinforce any weak spots and adjust the placement of branches or foliage as needed. Ensure that the entrance of the shelter is wide enough for you to comfortably enter and exit.

7. Test the Shelter

Before settling into your lean-to for the night, test its stability and waterproofing by gently pushing on the frame and checking for any leaks or weaknesses. Make any necessary adjustments to strengthen the structure and improve its weatherproofing.

8. Make Yourself Comfortable

Once your lean-to shelter is complete and secure, gather any additional supplies you may need for the night, such as bedding or extra clothing. Arrange your belongings inside the shelter to create a comfortable and functional living space.

By following these steps, you can build a sturdy and reliable lean-to shelter to protect yourself from the elements in a no-grid survival situation. With proper construction and maintenance, your lean-to will provide essential shelter and increase your chances of survival until help arrives.

Crafting a Debris Hut Shelter

Crafting a debris hut shelter is a more robust option for prolonged survival in a no-grid situation. It offers better protection from the elements and requires more effort but can be highly effective. Here's a step-by-step guide:

Materials Needed:

- Long, sturdy branches for the main frame
- Smaller branches, twigs, and debris for building walls and insulation
- Leaves, ferns, grass, or other foliage for covering
- Cordage, vines, or strips of bark for lashing (optional but helpful)

Steps:

1. Select a Location

Choose a flat area with good drainage, away from potential hazards like falling branches or flooding. Look for a spot where you can easily find materials for construction and where the prevailing wind won't blow directly into your shelter.

2. Construct the Frame

Begin by leaning two long branches against each other to create an A-frame structure. The branches should be sturdy enough to support the weight of the shelter. If necessary, secure the tops of the branches together using cordage or by weaving vines or strips of bark around them.

3. Build the Walls

Gather smaller branches, twigs, and debris to fill in the walls of your shelter. Lean them against the A-frame structure at an angle, starting from the ground and working your way up. Pack the debris tightly to create a sturdy wall that will provide insulation and protection from the elements.

4. Cover with Foliage

Once the walls are built, cover the entire structure with leaves, ferns, grass, or other foliage. Layer the foliage thickly to create a waterproof barrier that will shed rain and snow. Ensure that there are no gaps or holes where water can seep through.

5. Insulate the Interior

If you have additional foliage or debris available, use it to insulate the interior of your shelter. Layer leaves, grass, or other materials on the floor of the shelter to create a soft, insulating barrier between you and the ground. This will help retain body heat and keep you warm throughout the night.

6. Create the Entrance

Leave a small opening at one end of the shelter to serve as the entrance. Make sure it's large enough for you to crawl through comfortably but small enough to retain heat and keep out wind and rain. You can use additional branches or debris to partially block the entrance and provide extra protection.

7. Reinforce and Test

Once your debris hut shelter is constructed, take a moment to reinforce any weak spots and ensure that it's secure and stable. Test the shelter by gently pushing on the walls and roof to check for any signs of weakness or instability. Make any necessary adjustments to strengthen the structure and improve its durability.

8. Make Yourself Comfortable

Before settling into your shelter for the night, gather any additional supplies you may need, such as bedding, clothing, or food. Arrange your belongings inside the shelter to create a comfortable and functional living space. Finally, crawl inside and enjoy the warmth and security of your debris hut shelter.

By following these steps, you can craft a sturdy and effective debris hut shelter to protect yourself from the elements and increase your chances of survival in a no-grid situation. With proper construction and maintenance, your shelter will provide essential protection and comfort until help arrives.

Constructing a Raised Bed Shelter

Constructing a raised bed shelter in a no-grid survival situation can provide protection from the elements and potential hazards. Here's a step-by-step guide:

1. Select a Location

Choose a flat area with good drainage, preferably on higher ground to avoid flooding. Look for a spot with natural materials nearby that can be used for construction.

2. Gather Materials

Collect materials for the raised bed shelter. This may include sturdy branches, logs, sticks, rocks, leaves, grass, and any other natural resources available in the area.

3. Frame the Bed

Use long branches or logs to create the frame of the raised bed. Lay them out in a rectangular shape, securing the corners together with rope, vines, or other flexible materials. Ensure that the frame is sturdy and can support your weight.

4. Fill the Bed

Fill the frame with layers of natural materials such as leaves, grass, straw, or pine needles. These materials will provide insulation and cushioning.

5. Create Walls

Use additional branches or logs to build low walls around the raised bed, leaving an entrance for easy access. Stack the materials securely to form a barrier against wind and rain.

6. Cover the Shelter

Cover the top of the raised bed shelter with a waterproof material such as a tarp, poncho, or large leaves. Secure the cover tightly to prevent it from blowing away.

7. Insulate the Shelter

Add additional layers of natural materials to the walls and roof of the shelter to improve insulation and provide extra protection from the elements.

8. Add Finishing Touches

Once the basic structure is complete, add any additional features you may need, such as a door flap, ventilation holes, or a raised platform for storage.

9. Test the Shelter

Before relying on the shelter for an extended period, test its stability and effectiveness. Make any necessary adjustments to ensure it provides adequate protection and comfort.

10. Maintain the Shelter

Regularly inspect and maintain the raised bed shelter to ensure it remains secure and weather proof. Replace any damaged or worn materials as needed.

By following these steps, you can construct a raised bed shelter that provides a comfortable and secure refuge in a no-grid survival situation.

In no-grid survival situations, the ability to construct emergency shelters using natural materials can mean the difference between comfort and exposure, safety and danger. By following the step-by-step instructions outlined in this guide and selecting safe and suitable shelter locations, you can increase your chances of weathering the storm and emerging unscathed from even the most challenging wilderness environments.

BOOK 5:
SELF-SUFFICIENT FOOD PRODUCTION AND PRESERVATION

I. Sustainable Gardening Techniques

Sustainable gardening has become an increasingly popular practice in a society that values both freedom and environmental stewardship. This approach to gardening goes beyond simply planting seeds, as it encompasses a holistic understanding of its many facets. It is founded on respect for the environment, conservation, and the pursuit of food production that harmonizes with nature to weave a beautiful tapestry. Sustainable gardening embodies the principles of permanence, where planting signifies not just the anticipation of a harvest but also a commitment to caring for the planet.

Sustainable gardening is a comprehensive approach to creating and maintaining a garden ecosystem in harmony with the natural world. It's guided by a philosophy that promotes balance among plants, soil, water, and the broader environment, rather than a mere set of actions. The aim is not only to yield a bountiful harvest but also to cultivate a regenerative garden that can sustain itself and contribute positively to the surrounding ecosystem.

Soil is the cornerstone of any garden and a crucial element of sustainable gardening practices. Healthy soil teems with a diverse array of bacteria, fungi, and beneficial insects. Soil fertility is cultivated and preserved through practices such as using cover crops, composting, and minimizing soil disturbance. Successful plants and enduring gardens are the results of gardeners' ongoing efforts to nurture and enhance the soil.

An essential aspect of sustainable gardening is the reduction of water consumption. Strategies to conserve water and minimize waste include mulching, drip irrigation, and rainwater harvesting. Sustainable gardeners bear a dual responsibility: they aim to lessen their environmental impact and create gardens that can adapt to fluctuations in water availability by managing water resources efficiently.

Companion planting is a natural method of pest control and a fundamental component of sustainable gardening, aiming to work in harmony with nature. Certain plants emit chemicals that deter pests, while others attract beneficial insects, contributing to a healthy garden ecology. By minimizing the use of chemical pesticides, this approach promotes a more hospitable environment for both flora and fauna.

The extensive cultivation of a single crop, known as monoculture, can disrupt ecosystems and make plants more susceptible to pests and diseases. Sustainable gardeners advocate for crop diversity as a means to enhance soil health through varied root systems and to develop a resilient garden that can adapt to changing conditions.

Permaculture, short for "permanent agriculture," is a design philosophy that incorporates sustainable gardening practices. It focuses on developing self-sustaining ecosystems modeled after natural patterns.

Permaculture principles guide sustainable gardeners in creating efficient and regenerative landscapes that maximize yield while minimizing environmental impact.

A cornerstone of sustainable gardening is the adoption of organic farming techniques. This includes transitioning to natural fertilizers and pesticides instead of synthetic ones. Composting food scraps, using organic mulch, and employing natural pest control methods enhance the garden's overall health and the quality of the harvest.

Sustainable gardeners closely monitor planting times and plan for future harvests. They strategically plant crops at specific times of the year to align with natural growth seasons and space them out to ensure continuous harvesting. This methodical approach ensures a year-round supply of fresh produce.

Sustainable gardening is not limited to large open spaces; it is equally effective in small urban areas. Community gardens, vertical gardens, and container plantings demonstrate how sustainable practices can be adapted to limited spaces. These methods enable people to grow their own food sustainably, even in compact environments.

Promoting biodiversity is a crucial aspect of sustainable gardening, extending beyond merely cultivating fruits and vegetables. Biodiversity can be supported by using native plant species, creating pollinator habitats, and avoiding genetically modified organisms (GMOs). These practices benefit the overall ecosystem.

Sustainable gardening transcends individual plots to become a collective endeavor within a community. Many sustainable gardeners engage in educational outreach, sharing their knowledge and enthusiasm with others. This involvement often takes the form of community gardens, garden tours, and workshops, all of which play a role in encouraging sustainable and independent living.

Sustainable gardening encompasses more than just food production; it also includes the cultivation of culinary and medicinal gardens. By growing edible plants, medicinal herbs, and flowers, gardeners can enjoy a variety of fresh, organic ingredients for cooking and health purposes right in their own backyard.

The final step in sustainable gardening is preserving the harvest. Traditional methods such as canning, fermenting, drying, or root cellaring are eco-conscious ways for gardeners to enjoy their harvest throughout the year. These preservation techniques align with the principles of self-sufficiency and reduce reliance on store-bought produce.

As we engage in sustainable gardening, we join a movement that recognizes the interdependence of all living things. It's about more than just growing food; it's about cultivating a culture that values the land and works to ensure its continued vitality. Every aspect of a sustainable garden, from improving soil quality to preserving crops, is integrated into a regenerative whole. The goal of this introduction to sustainable gardening is to inspire everyone to participate in creating an environmentally friendly, resilient, and self-sufficient world.

II. Organic Farming Practices

Organic farming practices stand as a testament to the long-standing relationship between humans and the land. In exploring the intricacies of organic farming, we uncover that it's more than just planting crops. It's akin to a dance with nature, where each decision—be it where to plant, what to enrich the soil with, or how to tenderly care for the crops—is part of a nurturing cycle.

Organic farming is a philosophy that eschews synthetic fertilizers, pesticides, herbicides, and genetically modified organisms (GMOs) in favor of soil health, biodiversity, and ecological balance. It draws inspiration from time-honored farming techniques, emphasizing a holistic approach to agriculture.

In organic farming, soil is revered as the foundation of all agricultural ecosystems. It's viewed as a vibrant, living entity teeming with microbes, fungi, and nutrients. Practices like cover cropping, green manuring, and composting are employed to nurture and sustain soil health.

Embracing the wisdom of our ancestors, crop rotation is pivotal in organic farming. Rotating a diverse array of crops across seasons enhances ecological balance, reduces susceptibility to pests and diseases, and optimizes nutrient uptake. This practice underscores the importance of variety in maintaining a healthy farm ecosystem.

Organic farming seeks a delicate equilibrium, understanding that natural controls and feedback loops are key to managing pests. Strategies such as introducing beneficial insects, companion planting, and trap cropping minimize reliance on synthetic chemicals. The goal is to maintain a balanced ecosystem, not merely to eradicate pests.

Biodiversity is a cornerstone of organic farming, extending beyond a simple buzzword. Ecosystems rich in diversity are more resilient and capable of self-regulation. Organic farmers cultivate a variety of crops, incorporate flowering plants, and preserve native species to foster pollinator-friendly environments. This deliberate promotion of biodiversity enhances the overall health of the farm.

Organic agricultural strategies fundamentally emphasize refraining from using artificial inputs. Synthetic fertilizers, herbicides, and insecticides are all avoided in this approach. Instead, organic farmers use natural alternatives like cover crops, compost, and manure to improve soil fertility and provide plants with essential nutrients. This reliance on organic inputs ensures the long-term health of the land.

The use of genetically modified organisms (GMOs) has sparked heated debate in modern farming. Organic farmers are staunchly opposed to using GMOs when it comes to seeds. They prefer heirloom and open-pollinated varieties to maintain agricultural genetic diversity and promote resilience to changing environmental conditions.

Organic farming extends its principles beyond immediate production methods. Its holistic approach to resource management includes water conservation, energy efficiency, and waste reduction. Practices such as agroforestry, energy-efficient technology, and rainwater harvesting contribute to the larger goal of reducing environmental impact.

The natural cycle of the seasons plays a significant role in organic farming. Organic farmers time planting and harvesting to coincide with natural cycles to maximize crop yields in an environmentally conscious manner. Additionally, there is a strong emphasis on supporting local markets, minimizing the environmental impact of long-distance transportation, and building stronger community ties.

Many farmers seek organic certification so that consumers can trust that their produce is genuinely organic. Certifying organizations establish rigorous standards covering aspects from seed supply to pest management to soil health. Farmers voluntarily undergo certification to uphold organic agricultural standards, which acts as a benchmark for quality and sustainability.

Regenerative agriculture is gaining momentum within the natural farming community, aiming to go beyond mere sustainability. This approach seeks to actively restore ecosystems through practices such as agroecology, no-till farming, and agroforestry. Its goals include enhancing soil quality and biodiversity, thereby contributing to a more resilient agricultural system.

Organic farming is more than a set of practices; it's a social movement. Many organic farmers engage with their communities through educational programs, farm tours, and farmers' markets. By fostering a stronger connection between consumers and their food, organic farmers promote a culture of sustainability.

Despite its environmental and health benefits, organic farming faces challenges such as labor-intensive methods, production uncertainty, and the difficulty of transitioning from conventional to organic practices. However, the rewards are significant. Organic farming supports thriving ecosystems, produces nutritious products, and instills a sense of responsibility towards the environment. Embracing organic farming signifies a deeper bond between humans and the planet, based on mutual care and a commitment to preserving the land for future generations.

Organic farming transcends the boundaries of the farm, embodying a lifestyle that respects the interdependence of all life forms. Sustainable agriculture practices, from planting seeds to managing pests and preserving soil, reflect this ethos. As we delve into the world of organic farming, we are invited to appreciate the harmony of natural cycles, the vitality of the soil, and the potential of an earth-friendly approach to agriculture.

III. Permaculture Design Principles

Permaculture design principles are essential components of sustainable living, found in both urban gardens and rural landscapes. By focusing on harmony with nature, permaculture goes beyond traditional gardening to create ecosystems where each element supports the whole. As we explore permaculture, we find that principles based on careful observation, innovation, and deep ecological understanding offer a dynamic and abundant approach to our interactions with the land.

Observation is crucial in permaculture. Practitioners study the land's natural processes before planting or altering the terrain. This involves understanding patterns of sunlight, water flow, wind direction, and interactions between flora and fauna. By attuning to these environmental subtleties, permaculturists can design systems that harmonize with nature rather than disrupt it.

Permaculture divides spaces into zones based on the needs of plants and animals and their frequency of use. Zone 1, the most intensively managed area, typically surrounds a residence and is home to vegetables and herbs that require regular attention. As you move towards Zone 5, human intervention decreases, allowing nature to take precedence. Zoning ensures efficient resource and energy utilization.

Diversity is celebrated in permaculture through the creation of guilds—groups of mutually supportive plants that offer an alternative to monoculture. This approach might include a mix of pest-repellent species, dynamic accumulators, and nitrogen-fixers. Polycultures, inspired by natural ecosystems, enhance soil fertility and ecosystem resilience.

Edges, where two ecosystems meet, are areas of great potential. Permaculture design leverages edge effects to create inclusive and productive environments. For example, the edge of a pond can support a diverse array of plant and animal life. By maximizing these transitional zones, permaculturists can increase biodiversity and productivity.

'Stacking functions' is a permaculture principle that aims to yield multiple benefits from a single design element. Consider fruit trees: they provide fruit, shade the ground, attract pollinators, and enhance soil quality. This layered approach ensures that each element in the design serves multiple purposes, optimizing resource use.

Permaculture prioritizes energy efficiency through strategic planning. By situating energy-intensive elements closer to the home, permaculturists minimize their environmental impact. This thoughtful arrangement ensures that routine tasks, such as collecting eggs or harvesting vegetables, require minimal effort and time.

Water management is a crucial aspect of permaculture design. Techniques like swales, rain gardens, and other on-site water collection and storage methods are employed. The goal is to increase water retention through landscape design, reducing the need for external irrigation and supporting plant growth even in dry conditions.

Animals play a vital role in building resilient ecosystems in permaculture. Chickens, for example, can be used for pest control and soil aeration, while bees enhance pollination. Integrating animals into the design promotes system health by mimicking natural patterns and fostering symbiotic relationships between animals and plants.

Permaculture principles extend to sustainable home design, incorporating natural building materials like cob, straw bale, and recycled materials for their low environmental impact. Integrating energy-efficient design principles into construction plans allows for passive energy capture and storage, resulting in livable and sustainable homes.

Waste is viewed as a resource to be repurposed rather than discarded. Practices like recycling organic materials, composting, and vermiculture (using worms for composting) are part of the closed-loop systems that permaculture aims to establish. By reusing waste as input for other elements, the need for external resources and the amount of waste produced are reduced.

Permaculture emphasizes the importance of slow and small solutions, recognizing that incremental changes can have a significant impact. This principle encourages thoughtful, manageable interventions that align with natural rhythms and scales, leading to more sustainable outcomes.

We should acknowledge that designs are not static but constantly evolving, embracing change, using feedback loops within their systems to adapt to shifting conditions in nature and human needs. This flexibility allows designs to remain relevant and effective over time.

Permaculture thrives when extended to communities beyond individual sites. Collaborative efforts and knowledge sharing in sustainable development are common practices. Sustainable agriculture classes, seed exchanges, and community gardens can foster a collective sense of responsibility to make environmentally conscious lifestyle choices.

IV. Water Conservation in Gardening

Water conservation is a critical aspect of sustainable gardening, allowing us to connect with nature while caring for our plants. By adopting water-saving techniques, we can live in harmony with the environment and contribute to the global effort to alleviate water scarcity.

The first step in water conservation is recognizing the importance of saving water in our gardens. By learning about the global water crisis, gardeners can appreciate the impact of their efforts and strive to make a difference.

Selecting plants that are well-suited to the local climate and soil conditions can significantly reduce water usage. Native and drought-tolerant plants often require less water, making them ideal choices for water-conscious gardeners.

Applying a layer of organic mulch, such as wood chips, straw, or compost, helps protect the soil from erosion, retain moisture, and suppress weed growth. Mulching not only conserves water but also improves soil health, supporting a vibrant garden ecosystem.

Efficient irrigation is key to conserving water in the garden. Techniques such as drip irrigation, soaker hoses, or watering directly at the base of plants ensure that water is delivered where it's needed most, reducing evaporation and waste.

Collecting rainwater in barrels or cisterns allows gardeners to utilize nature's gift. By using stored rainwater for irrigation during dry periods, we can lessen our reliance on conventional water sources and mitigate the environmental impact of over-extracting from rivers and lakes.

Soil's ability to retain water is directly linked to its health. Loamy soils that are well-aerated and rich in organic matter retain more water. To enhance water efficiency, it's crucial to maintain soil structure by adding compost, using cover crops, and avoiding excessive soil disturbance.

The timing of irrigation is crucial for water conservation. Watering in the early morning or late evening reduces evaporation losses, ensuring that more water reaches the plant roots. Watering from the soil level rather than from above further minimizes evaporation.

Aerating the soil improves water penetration and distribution within the root zone. Compacted soil can hinder water absorption, leading to runoff and wasted water. Techniques like core aeration can alleviate compaction, allowing for better moisture absorption and root growth. Well-aerated soil enables deep water penetration, reducing the need for frequent watering.

Continuous monitoring is essential in gardening, just as it is in any caregiving role. A proactive water conservation strategy involves checking soil moisture levels, looking for signs of over- or under-watering, and adjusting irrigation schedules based on weather conditions.

V. Educating and Engaging the Community:

Water conservation extends beyond individual gardens to encompass community-wide efforts. By educating and involving the community, water stewardship becomes a shared responsibility. Workshops, community gardens, and information-sharing platforms can raise awareness and equip individuals with the tools needed to contribute to responsible water use.

Sustainable garden design includes not only plant choices but also infrastructure that supports water conservation. Incorporating hardscaping materials that allow water to percolate into the soil, as well as permeable paths and rain gardens that capture and filter runoff, can enhance the water-saving capabilities of a garden.

Recycling greywater—the runoff from washing machines and baths—can reduce the demand on freshwater resources. By redirecting this water to gardens, greywater recycling systems can contribute to efficient water use. Properly designed systems are crucial for the effective use of greywater in irrigation.

Water conservation in gardening is more than a collection of techniques; it's a commitment to looking beyond our own backyards and contributing to the broader efforts of sustainable water management. Every drop of water saved in our gardens plays a part in the global water conservation picture. It's a call to action to connect with nature, cultivate beautiful and sustainable gardens, and demonstrate our dedication to preserving this vital resource.

VI. Food Preservation

As the seasons change, there is often an abundance of food that exceeds immediate needs. Food preservation acts as a guardian of bounty, extending the life of the harvest long after it has been gathered. This chapter explores the diverse world of food preservation, uncovering the time-honored techniques that transform perishable goods into long-lasting sustenance.

Canning

Canning is a method of preserving food in which the food contents are processed and sealed in an airtight container, providing a shelf life typically ranging from one to five years. The canning process involves several steps to ensure food safety and preservation. Initially, the food is prepared according to the type of canning being used, which may involve peeling, slicing, chopping, or cooking. The prepared food is then placed into clean, sterilized jars, leaving the correct amount of headspace to allow for expansion during the canning process. Air bubbles are removed from the jars by gently tapping them or running a non-metallic spatula around the inside of the jar to prevent air pockets that can affect the sealing process.

Lids are placed on the jars, and screw bands are tightened just until fingertip tight, allowing air to escape during the canning process while preventing contaminants from entering. The jars are then placed in a canner filled with water and heated to a specific temperature for a set amount of time, depending on the type of food and canning method used. There are two main methods of processing: water bath canning, suitable for high-acid foods like fruits, pickles, and tomato products, where jars are submerged in boiling water; and pressure canning, required for low-acid foods like vegetables, meats, and poultry, where jars are processed in a pressure canner at a specific pressure to achieve the high temperatures needed to kill harmful bacteria and spores.

After processing, jars are removed from the canner and left to cool undisturbed for 12-24 hours, during which a vacuum seal is formed, and the lid will make a popping sound when successfully sealed. Once cooled, the screw bands are removed, and the jars are checked for proper sealing. Sealed jars are stored in a cool, dark place until ready to use. Canning is a popular way to preserve seasonal fruits and vegetables, allowing them to be enjoyed year-round, and is also used for creating jams, jellies, pickles, and relishes. Proper canning techniques are essential to prevent the growth of harmful bacteria, such as

Clostridium botulinum, which can cause botulism, a potentially deadly foodborne illness, making it crucial to follow tested recipes and guidelines from reputable sources for safe canning.

Pickling

Pickling is a traditional food preservation method that extends the shelf life of various foods, primarily vegetables and fruits, by submerging them in an acidic solution or through fermentation. The process of pickling can be categorized into two main types: vinegar pickling and fermentation pickling.

In vinegar pickling, food items are soaked in a mixture of vinegar, water, and salt, often with additional spices and flavorings. The high acidity of the vinegar prevents the growth of harmful bacteria, thereby preserving the food. Common additions to the pickling solution include sugar, herbs like dill and bay leaves, and spices such as mustard seeds, garlic, and chili peppers. This method can quickly preserve foods, and the resulting pickles are typically ready to eat within a few days to weeks, depending on the recipe.

Pickling not only preserves the lifespan of foods but also enhances their nutritional value by introducing vitamins such as Vitamin K and probiotics (in the case of fermented pickles). The process can transform the texture and flavor, adding a unique depth and zest that can complement various dishes.

Drying

Drying food is an ancient method of preservation that effectively reduces the moisture content in food items, inhibiting the growth of spoilage-causing microorganisms and enzyme activity. To successfully dry food, start by selecting high-quality, fresh produce or meats that are free from blemishes or bruises, as the quality of the dried product is heavily dependent on the freshness of the initial ingredients.

Begin the preparation process by thoroughly washing fruits and vegetables. Slice them into even, thin pieces to ensure uniform drying. For fruits like apples, bananas, and pears, consider pretreatment methods such as blanching or dipping in lemon juice to prevent browning. When drying meats, particularly for making jerky, choose lean cuts to minimize the presence of fat, which can spoil during storage. Trim away any fat and cut the meat into consistent, thin strips.

There are several methods you can employ to dry your food:

Sun Drying works best in hot, dry climates with strong sunlight. Spread the food pieces on clean trays or screens, cover them with a protective net to keep insects away, and place them in a sunny area. Rotate the pieces regularly to achieve even drying, which can take several days depending on the conditions.

Oven Drying involves setting your oven at its lowest temperature, typically around 140°F or 60°C. Arrange the food on wire racks over baking sheets to allow good air circulation and leave the oven door slightly ajar to let moisture escape. This method requires regular checks and can take anywhere from 6 to 12 hours.

Using a Dehydrator is perhaps the most controlled method. Lay the food slices on dehydrator trays in a single layer and follow the manufacturer's temperature guidelines, usually set between 125°F and 135°F (52°C to 57°C). The duration of dehydration varies widely based on the type of food and slice thickness.

To determine if the food is adequately dried, check for a leathery or brittle texture. There should be no visible moisture, and once cooled, the pieces should not stick together. After drying, you might consider 'conditioning'—a process where dried fruits and vegetables are placed in a loosely covered container two-thirds full and shaken daily for about a week to even out moisture content and reduce the risk of mold growth.

For storage, pack the dried foods in airtight containers and keep them in a cool, dark place. Vacuum-sealed bags, jars, or plastic containers with tight-fitting lids are ideal. When stored properly, dried foods can last several months to a year.

Ensure even drying by maintaining consistent thickness in your slices, rotating trays, and turning food pieces periodically when using an oven or dehydrator. Always label storage containers with the drying date to monitor shelf life. Avoid over-drying to facilitate easier rehydration later.

Fermentation

Fermentation is a time-honored method of food preservation that has been utilized by cultures around the world for centuries. It involves the controlled microbial transformation of food components, such as sugars and starches, into acids, gases, or alcohol. This process not only extends the shelf life of food but also enhances its flavor, texture, and nutritional value.

At the heart of fermentation is the activity of beneficial microorganisms, primarily bacteria, yeasts, and molds. These microbes consume the natural sugars present in the food and produce lactic acid, acetic acid, or ethanol as byproducts. The acidic environment created by these byproducts inhibits the growth of harmful bacteria, effectively preserving the food.

Fermentation can be categorized into two main types: lactic acid fermentation and alcoholic fermentation. Lactic acid fermentation is commonly used in the preservation of vegetables, dairy products, and meats. In this process, lactic acid bacteria convert sugars into lactic acid, creating tangy, flavorful foods like sauerkraut, kimchi, yogurt, and salami. Alcoholic fermentation, on the other hand, is primarily used in the preservation of fruits and grains, where yeasts convert sugars into ethanol and carbon dioxide, resulting in products like wine, beer, and bread.

One of the key advantages of fermentation as a preservation method is its ability to enhance the nutritional profile of foods. Fermented foods are often richer in vitamins, particularly B vitamins, and they can also improve digestive health by promoting a balanced gut microbiota. Furthermore, the fermentation process can break down anti-nutrients, such as phytates, making minerals more bioavailable.

To ensure successful fermentation, it is essential to maintain the right conditions, including temperature, salinity, and pH levels. Traditional fermentation techniques often rely on the natural presence of beneficial microbes, while modern practices may involve the addition of specific starter cultures to ensure consistency and safety.

To ferment food, start by choosing your produce; almost any vegetable can be fermented, with popular choices including cabbage (for sauerkraut), cucumbers (for pickles), carrots, radishes, and peppers, though fruits can also be fermented. Wash and chop your chosen produce into uniform pieces, with the size and shape depending on your preference and the type of ferment you're making.

Next, create a brine by dissolving salt in water at a ratio of about 2-5% salt to water by weight, which equates to 20-50 grams of salt per liter of water. Some recipes may call for adding spices or herbs to the brine for additional flavor. Place your prepared vegetables or fruits into clean, sterilized jars, packing them tightly to minimize air pockets and leaving some space at the top of the jar. Pour the brine over the produce, ensuring that it is completely submerged, and use a fermentation weight or a clean, smaller jar to keep the produce submerged if necessary.

Seal the jar with a lid, but not too tightly if using a standard lid, as gases will need to escape during fermentation. Alternatively, use an airlock lid designed for fermentation to allow gases to escape while preventing outside air from entering. Store the jar at room temperature, away from direct sunlight, with ideal fermentation temperatures typically between 55°F (13°C) and 75°F (24°C).

Monitor your ferment daily to ensure that the produce remains submerged and to release any built-up gases if not using an airlock lid. You may notice bubbles, a sour smell, and a change in color, all of which are normal signs of fermentation. Start tasting your ferment after a few days to determine if it has reached your desired level of sourness and flavor. Fermentation time can vary from a few days to several weeks, depending on the temperature, ingredients, and your taste preference.

Once your ferment has reached your desired flavor, transfer it to the refrigerator to slow down the fermentation process. Fermented foods can last for several months when stored properly in the fridge. It's important to maintain cleanliness throughout the fermentation process to avoid contamination, so use clean utensils and jars, and wash your hands thoroughly before handling the ingredients. Fermentation is both an art and a science, so feel free to experiment with different ingredients and flavors to create your unique fermented delights!

Salt Curing

Salt curing is a traditional food preservation method that uses salt to draw moisture out of food, primarily meat and fish, thereby inhibiting the growth of microorganisms that cause spoilage and decay. This technique, which dates back thousands of years, leverages the osmotic pressure created by salt to remove water, creating an environment unsuitable for bacterial growth.

The process of salt curing can be conducted through dry curing or brining. In dry curing, food is covered with salt and often a mix of ingredients such as sugar, spices, and nitrates or nitrites, which add flavor and assist in preservation. The mixture is thoroughly rubbed over the food, which is then left to cure for a period depending on its size and the desired level of preservation. During this time, the salt draws out moisture, and the food gradually becomes dehydrated.

Brining, or wet curing, involves submerging food in a solution of salt water that may also contain spices, sugars, and curing agents. This method is particularly effective for curing large cuts of meat or whole fish, ensuring the salt penetrates evenly throughout the food. The duration of brining can vary from a few hours to several weeks, based on the thickness of the meat and the intended flavor and texture.

After curing, the food may be rinsed (especially in the case of dry curing) to remove excess salt, and then it often undergoes additional preservation steps such as smoking or air drying, which further enhance flavor and longevity.

Salt curing not only extends the shelf life of food but also enhances its flavor, resulting in a product that is often denser, saltier, and more flavorful than its fresh counterpart. It is widely appreciated in various cuisines around the world for its unique taste and texture contributions to dishes.

This method of preservation is highly valued not only for its effectiveness but also for its ability to create distinctive culinary products like prosciutto, salt-cured fish, corned beef, and various types of salami and charcuterie. Each of these foods showcases the remarkable capacity of salt to transform and preserve natural flavors in a way that has been cherished through the ages.

Smoking

Smoking food is a traditional preservation method that imparts flavor, enhances color, and extends the shelf life of various foods, primarily meats and fish. This technique involves exposing the food to smoke from burning or smoldering materials, usually wood. Depending on the desired result, different types of wood, such as hickory, mesquite, apple, or cherry, are used, each imparting a unique flavor to the food.

There are three major types of smoking.

Cold Smoking, where foods are smoked at temperatures between 20°C to 30°C (68°F to 86°F). This method doesn't cook the food but imparts a smoky flavor and helps reduce moisture content, which inhibits bacterial growth. Cold-smoked products include cheeses, smoked salmon, and some sausages.

Hot Smoking approach cooks the food while smoking, with temperatures ranging from 52°C to 80°C (125°F to 176°F). It's used for meats like ribs, pork, and poultry, producing a moist, flavorful product that is ready to eat.

Smoke Roasting, also known as "barbecuing" or "pit roasting," this involves higher temperatures than hot smoking, effectively roasting the food in a smoky environment. It's commonly used for whole turkeys, chickens, or large cuts of beef and pork.

The smoking process typically starts with curing or salting the food, which draws out moisture and helps preserve it. The food is then placed in a smoker, where it's exposed to smoke that is carefully controlled to maintain the right temperature and smoke intensity. The duration of smoking can vary from a few hours to several days, depending on the size and type of food, as well as the specific characteristics desired in the finished product.

Smoking preserves food by slowing down the decomposition process. The smoke contains various chemical compounds, such as formaldehyde and acetic acid, which have antimicrobial properties. These compounds help inhibit the growth of bacteria, yeasts, and molds. Additionally, the smoking process also reduces the moisture content of the food, further preventing microbial growth.

While smoking can add unique flavors and help preserve food, it's important to consume smoked products in moderation. Some compounds formed during the smoking process, like polycyclic aromatic hydrocarbons (PAHs) and certain nitrosamines, can pose health risks if ingested in large amounts over time.

Smoked foods have a distinct taste and are often used as delicacies or integral parts of various culinary traditions around the world. They can be served on their own, used as ingredients in recipes, or used to enhance the flavor of other dishes.

Smoking as a preservation technique not only enhances flavor but also contributes to food safety and longevity, making it a cherished culinary art form that combines tradition with technique.

Advanced Modified Atmosphere Packaging (MAP)

Advanced Modified Atmosphere Packaging (MAP) is a sophisticated food preservation technique that enhances the shelf life and maintains the quality of perishable goods by altering the atmospheric composition inside the packaging. This method is widely applied to products ranging from fresh produce and meats to dairy and ready-to-eat meals.

In Advanced MAP, the air inside the packaging is replaced with a carefully calibrated mixture of gases, typically carbon dioxide, nitrogen, and oxygen. The specific ratios of these gases vary depending on the type of food; carbon dioxide inhibits bacterial growth, nitrogen acts as a filler to prevent package collapse, and controlled oxygen levels help reduce oxidation and spoilage. The packaging materials used are crucial, as they must be high-barrier types that prevent the exchange of gases with the external environment, thus maintaining the stability of the modified atmosphere over time. Materials like films, trays, and bags are customized to the specific gas requirements and the physical characteristics of the food product.

Effective sealing technology is also key to ensuring that the modified atmosphere is maintained throughout the product's shelf life, with advanced techniques used to create airtight seals that protect against contamination and prevent the dilution of the gas mixture.

The benefits of using Advanced MAP include an extended shelf life by slowing down the respiration rate of fresh produce and inhibiting microbial growth, maintenance of product quality by preserving texture, appearance, and nutritional value without the need for preservatives, and reduced food spoilage. The method provides flexibility to be customized for different products, making it a versatile solution suitable for a variety of food types.

Applications of Advanced MAP span across several food categories. In fresh produce, it helps maintain freshness by controlling respiration rates and delaying ripening. For meat and poultry, it preserves color and texture by limiting oxygen exposure, thus reducing spoilage and bacterial growth. In dairy products, it can prevent mold growth and maintain desired humidity levels, while for bakery and snack foods, it prevents staling and maintains crunchiness.

VII. Root Cellars

Root cellars are nature's way of preserving freshness. These underground storage spaces are carved out of the earth and utilize the cool, moist conditions of the soil to store perishables without the need for electricity or modern cooling systems. Constructing a root cellar requires an understanding of architecture and environmental conditions. Typically located beneath or adjacent to homes, these structures use clay and other natural materials to maintain a constant temperature and humidity. A well-designed ventilation system ensures air exchange while keeping cooling efficient. Vegetables such as carrots, potatoes, and apples can maintain their freshness for extended periods in this microclimate, which mimics natural environmental conditions.

Designing a root cellar is a thoughtful process that considers the soil, terrain, and climate. A well-ventilated and humidity-controlled environment is crucial for its success. In cold climates, insulated doors can be used to keep out freezing air, while in hot climates, additional measures may be taken to prevent overheating. Ventilation is strategically arranged to allow free air circulation, preventing the buildup of ethylene gas. Thoughtful shelving and layout provide ample space for each fruit and vegetable, ensuring easy access and minimizing the risk of spoilage.

Maintaining the right humidity level in a root cellar is like conducting a silent orchestra. Unlike the dry environment of a conventional refrigerator, root cellars need a certain amount of moisture to function effectively. Natural soil moisture helps maintain adequate humidity, protecting stored produce from drying out. In arid regions, materials like sand or sawdust can be used to retain moisture. Conversely, in humid areas, adequate ventilation is essential to prevent mold growth. This delicate balance between soil and stored produce ensures longevity and quality.

A root cellar is more than a storage space; it's a testament to the preservation of the harvest. Root cellaring is an act of entrusting the earth to care for our food. The cool, underground environment allows carrots and apples to retain their crispness and firmness without sprouting. Beyond physical preservation, root cellars play a role in maintaining culinary traditions and promoting sustainability. Communities that practice root cellaring often share knowledge and pass down skills through generations. Relying on natural cooling techniques reduces the ecological footprint, offering a timeless model for sustainable living.

VIII. Foraging Edible Plants

In a world increasingly turning towards sustainable living and reconnecting with nature, foraging for edible wild plants has gained popularity. Beyond the thrill of exploring the outdoors, foraging offers a unique opportunity to discover a bounty of nutritious and medicinal treasures hidden in plain sight. This guide aims to shed light on the identification, harvesting, and preparation of wild edible plants, emphasizing their nutritional value and potential medicinal properties.

Identification:

Before setting out on a foraging adventure, it's crucial to familiarize yourself with the flora native to your region. Invest in a reputable field guide or seek guidance from experienced foragers to learn to identify edible plants accurately. Pay close attention to key features such as leaf shape, color, texture, and growth patterns. Remember, accurate identification is paramount to ensure your safety and enjoyment.

Harvesting:

When harvesting wild edible plants, adopt a sustainable approach to ensure their continued abundance. Only collect plants from areas where they are plentiful, leaving behind enough to support their natural growth and propagation. Use sharp scissors or a knife to harvest plants cleanly, minimizing damage to the surrounding environment. Be respectful of wildlife habitats and avoid disturbing fragile ecosystems.

Preparation:

Once you've gathered your wild edibles, proper preparation is essential to unlock their culinary and medicinal potential. Begin by thoroughly washing the plants to remove any dirt or debris. Depending on the plant, you may need to remove tough stems, prickly thorns, or bitter components. Experiment with different cooking methods such as sautéing, steaming, or incorporating them into salads and soups to discover your preferred flavors and textures.

Nutritional Value:

Wild edible plants are nutritional powerhouses, often boasting higher nutrient levels than their cultivated counterparts. Rich in vitamins, minerals, and antioxidants, they offer a diverse array of health benefits. For example, dandelion greens are packed with vitamins A, C, and K, while stinging nettle is a potent source of iron and calcium. Incorporating a variety of wild edibles into your diet can enhance your overall health and well-being.

Medicinal Properties:

Beyond their nutritional value, many wild edible plants have been used for centuries in traditional medicine to treat various ailments. For instance, elderberries are prized for their immune-boosting properties, while plantain leaves possess anti-inflammatory and wound-healing abilities. However, it's essential to approach the medicinal use of wild plants with caution and seek guidance from qualified herbalists or healthcare professionals.

KNOWN NAME	LATIN NAME	DESCRIPTION	NUTRITION VALUE
1. **False morels**	Morchella	False morels are edible mushrooms with a black spore print. They are similar in taste to morel mushrooms but slightly sweeter. They are most commonly found in wooded areas, although they can be eaten year round.	False morels are not very nutritious because there is little protein and few nutrients in them, but they do contain some minerals such as calcium, iron, phosphorus, and zinc.
2. **Mulberry leaves**	Morus alba	Mulberry leaves are a common sight in fall. They have a light green color and a serrated edge. Mulberry leaves go well with all types of foods, especially rice porridge and gelatine. They can also be dried and powdered before being consumed.	Mulberry leaves have a high nutritional value because they contain insoluble dietary fibers, vitamin A, calcium, iron, and so on.

3. **Wild blackberries**	Rubus fruticosus	Wild blackberries are native to the Northern Hemisphere. They are pinkish red, with short stems and edible pulp.	Wild blackberries contain high amounts of vitamin C, which is an essential vitamin for good health. It is also very rich in water-soluble vitamins such as B1, B2 and B6.
4. **Raspberry Shoots and leaves**	Rubus occidentalis	The shoots can be harvested in the spring while they are still young, tender, and sweet. Raspberry leaves are used to make tea because they contain tannins, which are known for their medicinal properties.	They are particularly notable for their rich vitamin content. For instance, they are abundant in vitamin C, a crucial nutrient for overall health. Additionally, they are packed with water-soluble vitamins like B1, B2, and B6, further enhancing their nutritional profile.
5. **Wild ginger**	Asarum canadense	Wild ginger is widespread throughout North America. It can grow as a perennial or as an annual. The leaves and the roots of the wild ginger plant are edible.	They contain a lot of vitamin C, as well as calcium, phosphorus, and iron. Wild ginger is used to make tea or dried and powdered before consumption.
6. **Chickweed**	Stellaria media	Chickweed is a perennial that grows throughout the US. It has light green, triangular blades with long stems. The leaves are edible and can be consumed raw,	Chickweed is sometimes used as a spring vegetable or in salads, but it is more nutrient dense than other vegetables because it contains more iron than spinach and twice as

		cooked, or dried for later use.	much calcium as milk.
7. Shiitake Mushrooms	Lentinula edodes	The shiitake is a bracket mushroom that grows in clumps. It is brown to dark brown, with a thin stem that has rings (or partial rings) of gills attached to it. The caps are thin and flat with edges that slightly curl inward. This mushroom is most common in the fall and winter on oak trees.	Shiitake mushrooms also contain various other vitamins and minerals in smaller amounts, making them a nutritious addition to your diet. They are also valued for their potential health benefits, including immune system support and potential anti-inflammatory properties. Always ensure mushrooms are thoroughly cooked before consumption to maximize their digestibility and nutrient availability.
8. Portobello Mushrooms	Agaricus bisporus	The portobello mushroom is a pore mushroom with a white cap and a brown gill. It grows in groups on logs, stumps, and dead trees. This mushroom is most common in the fall.	Portobello mushrooms are also low in calories and fat while being rich in protein, making them a popular choice for those seeking a meat substitute. They can be grilled, sautéed, or stuffed, offering versatility in various dishes while providing a healthy dose of essential nutrients.

9. Milk Mushroom	Lactarius deliciosus	The milk mushroom is a pore mushroom with a small, white cap that grows in groups on rotting logs. They grow during the fall and winter on moderately decayed wood. It is edible when young but grows best in fall and winter.	Like many mushrooms, milk mushrooms are low in calories and fat, making them a nutritious addition to a balanced diet, especially for those watching their calorie and fat intake.
10. Honey Agaric	Armillaria mellea	The honey mushroom is a pore mushroom with a down-turned cap that resembles the head of an old man. It grows in groups on dead trees especially on damp forest floors, but it can be found in other places. This mushroom is edible when very young but grows best in fall and winter.	These mushrooms contain minerals like potassium, phosphorus, and selenium, which play essential roles in various bodily functions such as nerve function, bone health, and antioxidant defense.

11. Porcini Mushrooms 	Boletus edulis	The porcini mushroom is a pore mushroom with a brown cap and stem. The cap is beige when young and becomes tan-colored with age. It grows most commonly in fall, but it can grow in other seasons as well. It is edible when very young but grows best when aged for two months or more.	These mushrooms are also low in sodium and cholesterol, making them a healthy addition to various dishes. Additionally, porcini mushrooms contain certain compounds like beta-glucans, which have been associated with immune-boosting properties and may have anti-inflammatory effects.
12. Oyster Mushrooms 	Pleurotus ostreatus	The oyster mushroom is a pore mushroom with a cap shaped like an oyster shell. It grows in clusters on dead trees. The stems are white, thick, and somewhat twisted. This mushroom grows best in fall and winter on decayed wood.	oyster mushrooms are low in sodium and cholesterol, making them a healthy addition to various dishes. They also contain certain bioactive compounds such as beta-glucans, which have been associated with immune-boosting properties and may have anti-inflammatory effects.

Foraging for edible wild plants is a rewarding journey that connects us to the natural world and enriches our culinary and medicinal practices. By honing our skills in identification, harvesting, and preparation, we can enjoy the diverse flavors and health benefits that nature has to offer while fostering a deeper appreciation for the environment. So, venture outdoors with an open mind and a spirit of exploration and let the wonders of wild edibles nourish both body and soul.

IX. Water Bath Canning and Pressure Canning

What is Water Bath Canning?

Water bath canning is a method used to preserve high-acid foods like fruits, jams, jellies, and pickles. It involves submerging sealed jars of food in boiling water for a specific period, killing bacteria and sealing the jars to prevent spoilage.

Steps for Water Bath Canning:

1. Prepare Your Equipment: Gather canning jars, lids, rings, a canning pot or large stockpot, jar lifter, and canning funnel.
2. Prepare the Food: Wash and prepare your fruits or vegetables according to your recipe. Fill the jars with the prepared food.
3. Sterilize Jars: Place jars in boiling water for 10 minutes to sterilize. Keep them hot until ready for use.
4. Fill Jars: Using a canning funnel, fill jars with prepared food, leaving the recommended headspace. Remove air bubbles with a non-metallic spatula.
5. Seal Jars: Wipe jar rims with a clean, damp cloth. Place lids on jars and screw on rings until fingertip tight.
6. Process Jars: Lower filled jars into a pot of boiling water using a jar lifter, ensuring jars are covered with at least 1 inch of water. Process for the recommended time.
7. Remove Jars: Carefully remove jars from the water bath using a jar lifter and place them on a towel to cool.
8. Check Seals: After cooling, check that lids have sealed by pressing down on the center of each lid. If the lid doesn't move, it's sealed.

Tips for No-Grid Survival:

- Use a portable propane stove or open fire for boiling water.
- Opt for reusable canning jars and lids to minimize waste.
- Store canned goods in a cool, dark place away from sunlight to prolong shelf life.

Here are some recipes for your Water Bath Canning

1. **Pickled Beets**

Preparation Time: 85 minutes

Cooking Time: 35 minutes

Servings: 4 pints

Ingredients:

- 3 lbs. fresh, small beets
- 2 sugar cups
- 2 water cups
- 2 cider vinegar cups
- 2 cinnamon sticks
- 1 tsp. whole cloves
- 1 tsp. whole allspice

Directions:

1. Scrub beets and detruncate tops to 1 inch. Put in a Dutch oven and cover with water. Bring to a boil.
2. Reduce heat and let simmer, covered, until tender, 25-35 minutes.
3. Remove from water and let cool. Peel beets and cut into fourths.
4. Place beets in a Dutch oven with vinegar, sugar, and water.
5. Wrap cinnamon sticks, cloves, and allspice in a double thickness of cheesecloth. Add to beet mixture.
6. Bring to a boil, then reduce heat and cover. Let simmer 10 minutes. Discard spice bag.
7. Pack beets into four hot sterilized 1-pint jars to within 1/2 inch of the top.
8. Carefully scoop the hot liquid over beets, leaving 1/4 inch space of the top. Remove air bubbles and if necessary, adjust headspace by adding hot mixture. Wipe the rims carefully. Place tops on jars and screw on bands until fingertip tight.
9. Place jars into canner with boiling water, ensuring that they are completely covered with water. Let boil for 35 minutes. Remove jars and cool.

Nutrition:

Carbohydrates: 12 g

Fat: 0 g

Protein: 1 g

Sodium: 44 mg

Cholesterol: 0mg

Calories: 53

Sugars: 95.68g

2. **Pickled Brussels sprouts**

Preparation Time: 30 minutes

Cooking Time: 10 minutes

Servings: 6 pints

Ingredients:

- 3 lbs. fresh Brussels sprouts halved
- 1 medium sweet red pepper, finely chopped
- 6 garlic cloves, halved
- 1 medium onion, thinly sliced
- 2 tsp. crushed red pepper flakes
- 1 tbsp. celery seed
- 1 tbsp. whole peppercorns
- 3 tbsp. canning salt
- 1/2 sugar cup
- 21/2 white vinegar cups
- 21/2 water cups

Directions:

1. Fill a Dutch oven three-fourths full with water; bring to a boil.
2. Add Brussels sprouts in batches, cooking, uncovered, 4 minutes until tender-crisp.
3. With a slotted spoon remove and drop into ice water. Drain and pat dry.
4. Pack Brussels sprouts into six hot 1-pint jars.
5. Divide garlic and pepper flakes among jars.
6. In a large saucepan, bring remaining ingredients to a boil.
7. Carefully scoop the hot liquid over Brussels sprouts, leaving 1/4-inch space of the top. Remove air bubbles and if necessary, adjust headspace by adding hot mixture. Wipe the rims carefully. Place tops on jars and screw on bands until fingertip tight.
8. Place jars into canner with simmering water, ensuring that they are completely covered with water. Let boil for 10 minutes. Remove jars and cool.

Nutrition:

Carbohydrates 3g

Fat 0g;

Protein 1g;

Sodium 11mg

Cholesterol: 0mg

Calories: 14

3. Dill Pickle Spears

Preparation Time: 20 minutes

Cooking Time: 30 minutes

Servings: 7 pints

Ingredients

- 2 gallons water
- 1/2 cup pickling salt
- 8 pounds pickling cucumbers, quartered lengthwise
- 1/2 cup mustard seeds
- 24 fresh dill sprigs
- 6 garlic cloves, halved
- 1-1/2 quarts white vinegar
- 4 cups water
- 1/4 cup sugar
- 2 tbsps. Pickling spice

Directions

1. Combine the water and the salt in a large pot, stirring to dissolve the salt. Add the cucumbers and let it stand at room temperature for 12 hours.
2. Place 2 teaspoons of the mustard seeds, 2 of the dill sprigs, and 1 of the garlic cloves halves in each sterilized jar. Drain the cucumbers and divide them among the jars.
3. Mix together the vinegar, 4 cups of water, the sugar, and the pickling spice in a large saucepan. Bring the pot to a boil and cook for 10 minutes. Pour the hot vinegar mixture into the jars, filling to 1/2 inch from top. Clean the jar rims and add the lids to the jars. Process these in a boiling water bath for 20 minutes.

Nutrition:

Cholesterol 0mg

Calories 12 kcal

Fat: 0 g

Carbs: 1 g

4. **Garlic Dill Pickles**

Preparation Time: 20 minutes

Cooking Time: 15 minutes

Servings: 4 pints

Ingredients:

- 3 pounds Kirby cucumbers
- 1-1/2 c. apple cider vinegar
- 1 tsp. red chili flakes
- 2 tsps. Black peppercorns
- 4 tsps. Dill seed

- 8 peeled garlic cloves
- 2 tbsps. Pickling salt
- 1-1/2 c. water

Directions:

1. Wash and dry cucumbers, cutting them into spears. Remove the blossom end of cucumbers. In a saucepan combine vinegar, water and salt to make brine. Bring to boil over medium-high heat.
2. Equally divide the dill seed, garlic cloves, red chili flakes, and black peppercorns between the jars.
3. Pack cucumbers into the canning jars as tightly as you can without crushing them. Pour the brine over the cucumbers, filling jars to1/4 of an inch from top. Tap jars to help remove air bubbles from jars.
4. Wipe rims of jars and secure the lids in place.
5. Add jars to canning pot and boil for 15 minutes. Remove jars and place on towel on counter to cool at room temperature.
6. Once jars have cooled place in fridge. Let the pickles stay for at least one week before eating.

Nutrition:

Cholesterol 0mg

Calories 5 kcal

Fat: 0 g

Carbs: 12 g

Protein 0 g

Sugars: total10 g

5. **Mustard Pickled Vegetables**

Preparation Time: 20 minutes

Cooking Time: 15 minutes

Servings: 4 pints

Ingredients

- 1 head cauliflower
- 20 small green tomatoes
- 3 green bell peppers
- 4 cups pickling onions
- 24 2" pickling cucumbers
- 1 cup sugar
- 3/4 cup flour
- 1/2 cup dry mustard 1 tbsp. turmeric

- 7 cups apple cider vinegar
- 7 cups water
- 1 cup pickling (kosher) salt

Directions

1. Wash cauliflower and break into florets.
2. Wash tomatoes and quarter.
3. Wash peppers, cut in quarters, remove stem, seeds, and ribs.
4. Cut into 1/2-inch strips.
5. Peel onions.
6. Wash cucumbers, removing stem and blossom ends.
7. Toss vegetables in large non-reactive bowl or pot with salt.
8. Pour a quart of water over all, and let stand overnight.
9. Drain, cover with boiling water, and let stand ten minutes. Drain.
10. Combine sugar, flour, spices, vinegar, and 3 cups of water.
11. Cook until thick.
12. Add vegetables and continue cooking until vegetables are tender-crisp.
13. Pack into pint jars, dividing liquid evenly, and leaving 1/2-inch head space.
14. Wipe rims; screw on lids and rings.
15. Process jars in a boiling water bath for fifteen minutes.

Nutrition:

Cholesterol 0mg

Calories 10.1 kcal

Fat: 0 g

Carbs: 2 g

Protein 0 g

6. **Watermelon Pickles**

Preparation Time: 20 minutes

Cooking Time: 30 minutes

Servings: 4 pints

Ingredients

- 2 pounds watermelon rind
- 4 cups sugar
- 2 cups white vinegar
- 2 cups water
- 1 lemon, washed and sliced thinly
- 1 cinnamon stick
- 1 tbsp. whole cloves

Directions

1. Trim dark green and pink flesh from rind; cut into 1" cubes.
2. Combine 1/4 pickling salt and 1 quart of water.
3. Heat and stir until salt are dissolved.
4. Pour saltwater over rind cubes. Leave overnight.
5. Drain and rinse cubes.
6. Place in heavy pot or kettle.
7. Cover with cold water and cook until tender; drain.
8. Combine sugar, vinegar, water, lemon slices in a heavy pot.
9. Place cinnamon and cloves in a cheesecloth bag and put bag in vinegar mixture.
10. Simmer mixture 10 minutes and remove spice bag.
11. Add rind cubes to vinegar mixture and continue cooking until cubes are translucent.
12. Pour into hot, sterile, pint jars, dividing syrup evenly, and leaving 1/2 inch head space.
13. Process jars in a boiling water bath for fifteen minutes.

Nutrition:

Cholesterol 0mg

Calories 70 kcal

Fat: 0 g

Carbs: 17 g

Protein 0 g

7. Mango Pineapple Salsa

Preparation Time: 10 minutes

Cooking Time: 30 minutes

Servings: 4

Ingredients:

- 2 mangoes, peeled and chopped
- 2 jalapenos, chopped
- 1 sweet pepper, chopped
- 1 onion, chopped
- 2 garlic cloves, minced
- 1 tsp. ginger, grated
- 1/4 cup vinegar
- 1/4 cup lime juice
- 1/3 cup sugar
- 3 cups pineapple, chopped
- 1 1/2 lbs. tomatoes, cored and chopped
- 1/2 tsp. salt

Directions:

1. Add all ingredients into the large pot and bring to boil.
2. Reduce heat and simmer for 10 minutes. Stir frequently.
3. Remove pot from heat. Ladle salsa into the clean jars. Leave 1/2-inch headspace.
4. Seal jar with lids. Process in a water bath canner for 20 minutes.
5. Remove jars from the water bath and let it cool completely.
6. Check seals of jars. Label and store.

Nutrition:

Cholesterol 0mg

Calories 280

Fat 1 g

Carbohydrates 70 g

Sugar 60 g

Protein 4 g

8. Chipotle BBQ Sauce

Preparation Time: 15 minutes

Cooking Time: 50 minutes

Servings: 48

Ingredients

- 1 tablespoon olive oil
- 1/4 cup onion, chopped finely
- 2 garlic cloves, minced
- 2 cups tomato sauce
- 11/2 (12-ounce) cans tomato paste
- 13/4 ounces canned chipotle peppers in adobo sauce
- 1 cup apple cider vinegar
- 1/2 cup honey
- 1/2 cup brown sugar
- 1 teaspoon dry mustard
- 1/2 teaspoons pickling salt
- 1/2 teaspoon ground black pepper

Directions:

1. In a nonreactive saucepan, heat olive oil over medium heat and sauté the onion and garlic for about 2–3 minutes.
2. Add in the remaining ingredients and cook until boiling.

3. Now set the heat to low and cook for about 15–20 minutes, stirring occasionally.
4. Remove the saucepan of sauce from heat and with an immersion blender, blend until smooth.
5. Return the pan over low heat and cook for about 20–25 minutes, stirring occasionally.
6. In 6 (1/2-pint) hot sterilized jars, divide the sauce, leaving about 1/2-inch space from the top.
7. Slide a small knife around the insides of each jar to remove air bubbles.
8. Wipe any trace of food off the rims of jars with a clean, moist kitchen towel.
9. Close each jar with a lid and screw on the ring.
10. Arrange the jars in a boiling water canner and process for about 20 minutes.
11. Remove the jars from water canner and place onto a wood surface several inches apart to cool completely.
12. After cooling with your finger, press the top of each jar's lid to ensure that the seal is tight.
13. The canned sauce can be stored in the refrigerator for up to 1 year.

Nutrition:

Cholesterol 0mg

Calories 32

Total Fat 0.4 g

Carbs 7.2 g

Protein 0.7 g

9. Peppers & Tomato Salsa

Preparation Time: 15 minutes

Cooking Time: 15 minutes

Servings: 48

Ingredients

- 10 cups tomatoes; peeled, cored, and chopped
- 5 cups onions, chopped
- 5 cups green bell peppers, seeded and chopped
- 21/2 cups jalapeño peppers, seeded and chopped
- 3 garlic cloves, chopped finely
- 2 tablespoons fresh cilantro, chopped finely
- 11/4 cups cider vinegar
- 1 tablespoon salt

Directions:

1. In a nonreactive saucepan, add all ingredients over medium-high heat and cook until boiling, stirring continuously.
2. Now set the heat to low and cook for about 10 minutes, stirring frequently.
3. In 6 (1-pint) hot sterilized jars, divide the salsa, leaving about 1/2-inch space from the top.
4. Slide a small knife around the insides of each jar to remove air bubbles.

5. Wipe any trace of food off the rims of jars with a clean, moist kitchen towel.
6. Close each jar with a lid and screw on the ring.
7. Arrange the jars in a boiling water canner and process for about 15 minutes.
8. Remove the jars from water canner and place onto a wood surface several inches apart to cool completely.
9. After cooling with your finger, press the top of each jar's lid to ensure that the seal is tight.
10. The canned salsa can be stored in the refrigerator for up to 1 month.

Nutrition:

Cholesterol 0mg

Calories 19

Total Fat 0.2 g

Sodium 241 mg

Carbs 3.9 g

Protein 0.7 g

10. Pear Caramel Sauce

Preparation Time: 15 minutes

Cooking Time: 30 minutes

Servings: 32

Ingredients

- 2 pounds ripe pears, cored and cut into pieces
- 2 teaspoons vanilla bean paste
- 1 teaspoon sea salt
- 13/4 cups water, divided
- 3 cups granulated sugar

Directions:

1. In a blender, add chopped pears, vanilla bean paste, salt, and 1/4 cup of water and pulse until smooth.
2. Transfer the pear puree into a bowl and set aside.
3. In a heavy-bottomed saucepan, add sugar and remaining water over medium-high heat and simmer for about 15–20 minutes, swirling the pan often.
4. Remove the saucepan of sugar syrup from heat and stir in the pear puree.
5. Return the saucepan over medium-low heat and cook for about 5–10 minutes or until the temperature of caramel sauce reaches between 215°F–225°F, stirring continuously.
6. In 4 (1/2-pint) hot sterilized jars, divide the sauce, leaving about 1/2-inch space from the top.
7. Slide a small knife around the insides of each jar to remove air bubbles.
8. Wipe any trace of food off the rims of jars with a clean, moist kitchen towel.

9. Close each jar with a lid and screw on the ring.
10. Arrange the jars in a boiling water canner and process for about 10 minutes.
11. Remove the jars from water canner and place onto a wood surface several inches apart to cool completely.
12. After cooling with your finger, press the top of each jar's lid to ensure that the seal is tight.
13. The canned sauce can be stored in the refrigerator for up to 1 year.

Nutrition:

Cholesterol 0mg

Calories 87

Total Fat 0 g

Sodium 58 mg

Carbs 23.1 g

Protein 0.1 g

11. Green Tomato Jam

Preparation Time: 10 minutes

Cooking Time: 20 minutes

Servings: 3 half-pints

Ingredients:

- 21/2 cups pureed green tomatoes
- 2 cups stevia
- 1 package raspberry gelatin

Directions:

1. In a large saucepan, bring stevia and tomatoes to a boil.
2. Reduce heat and let simmer, uncovered, for 20 minutes.
3. Remove from the heat and add gelatin, stirring until dissolved.
4. Skim off any foam.
5. Scoop the hot mixture in hot sterilized half-pint jars, leaving 1/4-inch space of the top. Remove air bubbles and if necessary, adjust headspace by adding hot mixture. Wipe the rims carefully. Let cool before covering with lids. Refrigerate up to 3 weeks.

Nutrition:

Carbohydrates 20 g

Fat 0 g

Protein 1 g

Sodium 10 mg

Cholesterol 0mg

Calories 81

12. Apricot Amaretto Jam

Preparation Time: 30 minutes

Cooking Time: 10 minutes

Servings: 8 half-pints

Ingredients:

- 4 1/4 cups peeled, crushed apricots
- 1/4 cup lemon juice
- 6 1/4 cups erythritol, divided
- 1 package powdered fruit pectin
- 1/2 teaspoon unsalted butter
- 1/3 Cup amaretto

Directions:

1. In a Dutch oven, combine lemon juice and apricots.
2. In a small bowl, combine pectin and 1/4 cup erythritol. Stir into apricot mixture and add butter. Bring to a full boil over medium-high heat, stirring constantly.
3. Stir in the remaining erythritol and let boil 1-2 minutes, stirring constantly.
4. Remove from heat and stir in amaretto.
5. Let the jam sit for 5 minutes, stirring occasionally.
6. Divide the hot mixture between eight hot sterilized half-pint jars, leaving 1/4-inch space of the top. Wipe the rims carefully. Place tops on jars and screw on bands until fingertip tight.
7. Place jars into canner with boiling water, ensuring that they are completely covered with water. Let boil for 10 minutes. Remove jars and cool.

Nutrition:

Carbohydrates 21 g

Fat 0 g

Protein 0 g

Cholesterol

Sugar 0 g 0mg

Calories 86

13. Blueberry Cinnamon Jam

Preparation Time: 35 minutes

Cooking Time: 10 minutes per batch

Servings: 9 half-pints

Ingredients:

- 8 cups fresh blueberries
- 6 cups erythritol
- 3 tbsp. lemon juice
- 2 tsp. ground cinnamon
- 2 tsp. grated lemon zest
- 1/2 tsp. ground nutmeg
- 2 (3 oz.) pouches liquid fruit pectin

Directions:

1. Place blueberries in a food processor and process until well blended. Transfer to a stockpot.
2. Stir in the lemon juice, erythritol, cinnamon, nutmeg, and lemon zest. Bring to a rolling boil over high heat, stirring constantly.
3. Stir in pectin. Boil for 1 minute, stirring constantly.
4. Remove from the heat; skim off foam.
5. Scoop the hot mixture in hot sterilized half-pint jars, leaving 1/4-inch space of the top. Remove air bubbles. Wipe the rims carefully. Place tops on jars and screw on bands until fingertip tight.
6. Place jars into canner with boiling water, ensuring that they are completely covered with water. Let boil for 10 minutes. Remove jars and cool.

Nutrition:

Carbohydrates 19 g

Fat 0 g

Protein 0 g

Cholesterol 0mg

Calories 74

Sugar 0 g

14. Carrot Marmalade

Preparation Time: 10 minutes

Cooking Time: 40 minutes

Servings: 48

Ingredients:

- 2 cups grated carrots

- 2 1/2 cups stevia
- 2 cups water
- 1 orange
- 1 lemon

Directions:

1. Shred orange and lemon in a large saucepan.
2. Add remaining ingredients into the saucepan and bring to boil over medium heat.
3. Reduce heat to low and simmer for 30 minutes or until thickened.
4. Once marmalade is thickened then remove the pan from heat.
5. Ladle the marmalade into the clean and hot jars. Leave 1/2-inch headspace. Remove air bubbles.
6. Seal jars with lids and process in a boiling water bath for 5 minutes.
7. Remove jars from the water bath and let it cool completely.
8. Check seals of jars. Label and store.

Nutrition:

Cholesterol 0mg

Calories 43

Fat 0 g

Carbohydrates 11.3 g

Sugar 0 g

Protein 0.1 g

15. **Raspberry Peach Jam**

Preparation Time: 35 minutes

Cooking Time: 15 minutes

Servings: 3 half-pints

Ingredients:

- 2 2/3 cups peeled, chopped peaches
- 1 1/2 cup crushed raspberries
- 3 cups stevia
- 1 1/2 tsp. lemon juice

Directions:

1. In a Dutch oven, combine all ingredients.
2. Cook over medium-low heat. Stir until the stevia has dissolved and the mixture is bubbly about 10 minutes.
3. Bring to a full boil for 15 minutes, stirring constantly.
4. Remove from heat and skim off foam.

5. Carefully scoop the hot mixture into hot sterilized half-pint jars, leaving 1/4-inch space of the top. Remove air bubbles. Wipe the rims carefully. Place tops on jars and screw on bands until fingertip tight.
6. Place jars into canner with boiling water, ensuring that they are completely covered with water. Let boil for 15 minutes. Remove jars and cool.

Nutrition:

Carbohydrates 8g

Fat 0

Protein 0g

Cholesterol 0mg

Calories 33

Sugar 0 g

Pressure Canning:

What is Pressure Canning?

Pressure canning is a method used to preserve low-acid foods like vegetables, meats, and soups. These foods require higher temperatures than what can be achieved with water bath canning to kill bacteria and prevent spoilage.

Steps for Pressure Canning:

1. Prepare Your Equipment: Gather canning jars, lids, rings, a pressure canner, jar lifter, and canning funnel.
2. Prepare the Food: Wash and prepare your vegetables, meats, or soups according to your recipe. Fill the jars with the prepared food.
3. Sterilize Jars: Place jars in a warm oven or dishwasher to sterilize. Keep them hot until ready for use.
4. Fill Jars: Using a canning funnel, fill jars with prepared food, leaving the recommended headspace. Remove air bubbles with a non-metallic spatula.
5. Seal Jars: Wipe jar rims with a clean, damp cloth. Place lids on jars and screw on rings until fingertip tight.
6. Prepare Pressure Canner: Add the specified amount of water to the pressure canner and insert the rack. Place filled jars inside.
7. Secure Lid: Follow the manufacturer's instructions for sealing the pressure canner lid and bringing it to the correct pressure.
8. Process Jars: Process jars at the recommended pressure for the specified time, adjusting for altitude if necessary.
9. Cool and Check Seals: After processing, allow the pressure canner to cool naturally. Remove jars and check seals after cooling.

Tips for No-Grid Survival:

- Use a portable propane stove or open fire for pressure canning.
- Maintain a steady pressure throughout the canning process, adjusting heat as needed.
- Practice proper safety precautions when using a pressure canner to prevent accidents.
- With these methods, you can safely preserve food for long-term storage in a no-grid survival situation, ensuring you have access to nutritious meals when resources are limited.

1. Canned Chicken and Gravy

Preparation time: 25 minutes

Cooking time: 35 minutes

Servings: 4 or 5 quart jar

Ingredients:

- 1 cup chopped onion
- 1 cup chopped celery
- 1 cup diced potatoes
- 2 lbs. boneless chicken breasts
- 2 tsps. salt
- 2 tsps. poultry seasoning
- 4 tbsps. white wine
- enough chicken stock

Directions:

1. Sterilize the jars in a pressure canner as indicated in the general guidelines of this book. Allow the jars to cool.
2. Set all ingredients in a saucepan and allow to simmer for 10 minutes over medium high heat.
3. Put the chicken and vegetables into the jars. Pour over enough broth to cover the chicken. Leave a 1/2-inch headspace.
4. Remove the air bubbles and close the lid.
5. Set the jars in the pressure canner and process for 75 minutes.

Nutrition:

Calories: 562

Protein: 77.7g

Fat: 22.2g

Carbs: 7.1g

2. Pressure Canned Rosemary Chicken

Preparation time: 25 minutes

Cooking time: 1 hour

Servings: 8 to 10 quart jar

Ingredients:

- 20 sprigs of rosemary
- 10 lbs. boneless chicken breast
- 1/4 cup salt

Directions:

1. Add a sprig of rosemary to each sterilized jar.
2. Slice the chicken breasts into large chunks and pack in the jars leaving a 1.5-inch headspace.
3. Attach a sprig of rosemary at the top then add a tbsp. of salt in each jar.
4. Rinse the rims of the jar with a clean damp towel, and then place the lids and the rings. Transfer the jars to the pressure canner and process them at 10 pounds pressure for 75 minutes.
5. Wait for the pressure canner to depressurize to zero before removing the jars using cooking tongs.
6. Transfer the jars on a cooling rack for 24 hours to seal then store in a cool dry place.

Nutrition:

Calories: 182.6

Fat: 7.8g

Carbs: 1.0g

Protein: 18.8g

3. Asian-Style Sweet and Sour Chicken

Preparation Time: 15 minutes

Cooking Time: 15 minutes

Servings: 7 (1-quart) jars

Ingredients:

- 5 pounds(2.3 kg) boneless skinless chicken, cut into bite-sized pieces
- 3 cups pineapple juice
- 3/4 cup brown sugar
- 11/4 cups apple cider vinegar
- 6 tbsps soy sauce
- 4 tbsps tomato paste
- 1 tsp ground ginger
- 4 cloves garlic, minced
- 2 medium onions, diced
- 3 bell peppers, diced
- 1 whole pineapple, cleaned and diced
- crushed chili pepper, to taste (optional)

Directions:

1. Combine pineapple juice, sugar, vinegar, soy sauce, tomato paste, ginger, and garlic in a large saucepan, bring to a boil, stirring frequently. Reduce the heat and simmer until the sugar is dissolved and the mixture is smooth.
2. Layer chicken, onions, peppers, and pineapple in your jars. If you're using crushed chilis, add them now.
3. Ladle the sauce over the contents of the jars.
4. Wipe the rims of the jars, put the lids on, and process in a pressure canner at 11 PSI for 90 minutes, adjusting for altitude.

Nutrition:

Calories: 391

Fat: 3.5g

Carbs: 32.6g

Protein: 19.5g

4. Pineapple Chicken

Preparation time: 25 minutes

Cooking time: 1 hour and 30 minutes

Servings: 6 quart jar

Ingredients:

- 3 cups pineapple juice
- 3/4 cup brown sugar
- 1 1/4 cups apple cider vinegar
- 6 tbsps. soy sauce
- 4 tbsps. tomato paste
- 1 tsp. ground ginger
- 4 minced garlic cloves
- 5 lbs. chopped boneless and skinless chicken
- 2 diced onions
- 3 diced bell peppers
- 1 diced pineapple
- crushed chili pepper, to taste

Directions:

1. In a sizable saucepan, bring to a boil pineapple juice, sugar, vinegar, soy sauce, tomato paste, ginger, and garlic, stirring frequently.
2. Boil for the sugar to dissolve and until the mixture is smooth.
3. In your jars, layer chicken, onions, peppers, and pineapple. If you're using crushed chilis, add them now.

4. Put the sauce over the contents of the jars.
5. Wipe the rims of the jars, put the lids on, and process in a pressure canner at 11 PSI for 90 minutes, adjusting for altitude.

Nutrition:

Calories: 391

Fat: 3.5g

Carbs: 32.6g

Protein: 19.5g

5. Bourbon-Flavored Salmon

Preparation time: 25 minutes

Cook time: 1 hour and 40 minutes

Servings: 6 pint jars

Ingredients:

- 1 (3-pound / 1.4-kg) salmon fillet, skin on or skinned, cut crosswise into 3-inch pieces
- 1 cup orange marmalade
- 1/4 cup bourbon
- 2 tablespoons lemon juice

Directions:

1. This recipe is raw packed, so have clean jars resting in lukewarm, temperate water.
2. Pack the raw salmon tightly into jars leaving a generous 1-inch headspace. Do not cover the fish with water.
3. In a small saucepan, merge the orange marmalade, bourbon, and lemon juice. Set to a boil over medium-high heat, stirring constantly. Add 2 tablespoons of the marmalade mixture to each pint jar and 1 tablespoon to each half-pint.
4. Set the rim of each jar with a washcloth dipped in distilled white vinegar. Set a lid and ring on each jar and hand-tighten.
5. Set the pressure canner with 3 quarts of water and add 2 tbsp. distilled white vinegar. Set the jars in the pressure canner, lock the pressure canner lid, and bring to a boil over high heat. Let the canner vent for 10 minutes. Secure the vent and continue heating to achieve 11 PSI for a dial gauge and 10 PSI for a weighted gauge. Process both pints and half-pints for 100 minutes.
6. After the processing time has been reached, set off the heat. Set the pressure in the canner to reach zero on its own, which usually takes about 30 minutes. Safely remove the canner lid and allow jars to sit for 10 minutes inside canner before removing.

Nutrition:

Calories: 314

Sodium: 73mg

Dietary Fiber: 2.4g

Total Fat: 5.1g

Total Carbs: 1.3g

6. Pressure Canned Tilapia

Preparation Time: 15 minutes

Cooking time: 1 hour and 40 minutes

Servings: 5 pint jars

Ingredients:

- 5 lbs. tilapia fillets
- Canning salt
- Lemon juice
- 1 jalapeño pepper

Directions:

1. Place 1 slice of jalapeño pepper into each jar. Fill jars with fish to 1/2 inch from the top. Add 1/4-teaspoon of canning salt and 1 tsp lemon juice per pint.
2. Use a knife to jiggle the meat and remove any air pockets. Wipe rim of jar clean. Heat lids in hot water for 3 minutes; place lids on jars and tighten rings slightly.
3. Arrange the jars in the canner and fill with water up to the jar rings. Close and lock the pressure canner and bring to a boil over high heat, then add cooking weight to the top.
4. After 20 minutes, adjust the heat to medium and cook for 80 minutes. Turn off the heat and leave the canner alone until it has cooled completely to room temperature.
5. After canner has cooled, remove the jars from the canner and check for sealing. If jars have sealed, store for up to 2 years; if not, use the meat right away.

Nutrition:

Calories: 96

Fat: 1.7g

Carbs: 0g

Protein: 20.08g

7. Lamb Chunks

Preparation Time: 10 minutes

Cooking Time: 90 minutes

Servings: 1 pint jars

Ingredients:

- 1 tsp. salt (in each quart jar)
- 1 lb. lamb meat, cut into chunks
- meat broth, boiling/tomato juice/water

Directions:

1. Remove any excess fat from your chilled, high-quality meat. If using wild meats, soak first in brine water (5 quarts water plus 5 tbsps. salt) to remove the strong flavors. Discard the large bones after rinsing.
2. Roast or stew your meat chunks until rare (you may also brown them using a little bit of fat). Add the cooked meat to clean and hot Mason jars filled with salt (1 tsp.). Pour your preferred liquid into it, making sure to leave an inch of headspace.
3. If choosing to raw pack, fill each jar with 1 tsp. salt first before adding your raw meat chunks, letting an inch of headspace remain without adding any liquid.
4. Get rid of air bubbles and secure the lids on the jars. Process in the pressure canner for 1 hour and 15 minutes (pints) or 1 hour and 30 minutes (quarts).

Nutritional Info:

Calories: 104

Fat: 4.1g

Carbs: 16.3g

Protein: 1.3g

8. Pressure Canned Beef Short Rib

Preparation time: 60 minutes

Cooking time: 75 minutes

Servings: 12 half pint jars

Ingredients:

- 10 lb. Beef short rib
- Water
- Pickling salt

Directions:

1. Heat a skillet sprayed with cooking spray. Brown the ground beef and keep it covered in a bowl to keep it hot.
2. Pack the beef in sterilized jars leaving a 1-inch headspace. Add a 1/2 tablespoon of pickling salt in each jar.
3. Add boiling water or stock to each jar, then remove the bubbles.

4. Set the rims and place the lids on. Transfer the jars to the pressure canner and process them at 10 pounds for 75 minutes.
5. Wait for the pressure canner to depressurize to zero before removing the jars.
6. Set the jars on a cooling rack for 24 hours then store in a cool dry place.

Nutrition:

Calories: 205

Saturated fat 3.4g

Carbs: 0g

Protein: 28.9g

Sugars: 0g

9. **Pressure Canned Ground Beef**

Preparation time: 50 minutes

Cooking time: 75 minutes

Servings: 12 pint jars

- Ingredients:
- 12 lb. Ground beef
- Water
- salt

Directions:

1. Heat a skillet sprayed with cooking spray. Brown the ground and keep it in a covered bowl to keep them hot.
2. Pack the beef in sterilized jars leaving a 1-inch headspace. Add a 1/2 tablespoon of pickling salt in each jar.
3. Add boiling water or stock, then remove the bubbles.
4. Wipe the rims and set the lids on. Transfer the jars to the pressure canner and process them at 10 pounds for 75 minutes.
5. Wait for the pressure canner to depressurize to zero before removing the jars.
6. Set the jars on a cooling rack for 24 hours then store in a cool dry place.

Nutrition:

Calories: 124

Saturated fat: 1.8g

Carbs: 0g

Protein: 21.2g

Sugars: 0g

Sodium: 62mg

10. Pressure Canned Stewing Beef

Preparation time: 60 minutes

Cooking time: 75 minutes

Servings: 5 pint jars

Ingredients:

- 5 lb. Stewing beef
- Water
- Pickling salt

Directions:

1. Trim any gristle on the stewed beef then cut it into strips or into cubes
2. Heat a skillet sprayed with cooking spray. Brown the stewed beef in batches and keep it in a covered bowl to keep them hot.
3. Pack the beef in sterilized jars leaving 1-inch headspace. Add a 1/2 tablespoon of pickling salt in each jar.
4. Add boiling water or stock, then remove the bubbles.
5. Wipe the rims and set the lids on. Transfer the jars to the pressure canner and process them at 10 pounds for 75 minutes.
6. Wait for the pressure canner to depressurize to zero before removing the jars.
7. Set the jars on a cooling rack for 24 hours then store in a cool dry place.

Nutrition:

Calories: 186

Saturated fat: 2.4g

Carbs: 0g

Protein: 30.3g

Sugars: 0g

Sodium: 66mg

Potassium: 403mg

11. Beef Tips and Gravy

Preparation Time: 35 minutes

Cooking Time: 90 minutes

Servings: 12 pints

Ingredients:

- Tablespoons extra-virgin olive oil, divided
- 12 pounds beef stew meat
- 2 teaspoons coarse sea salt (optional)
- 1 teaspoon ground black pepper
- 2cups hot water, divided
- 1/2 cup Clear gel
- 12 garlic cloves

Directions:

1. In a skillet, warmth 1 tablespoon of olive oil. Working in small batches, attach the beef and lightly brown each side, about 3 minutes. Flavor each batch with a dash of sea salt and black pepper while in the skillet. Attach 1 additional tbsp. of oil while browning each batch. Set batches aside.
2. Keeping the skillet on medium-high heat, slowly attach 4 cups of hot water to the skillet. Set in the Clear Jel and slowly add the remaining hot water, continuing to whisk. Set to a boil for 2 minutes, then detach from the heat. Set the gravy aside.
3. Attach 2 whole garlic cloves to each warm quart and 1 to each warm pint. Using a spoon, raw pack the meat into jars, leaving a generous 1 inch of headspace. Set the hot gravy mixture over the meat, maintaining the headspace. Detach any air bubbles and add additional sauce if necessary to maintain the headspace.
4. Clean the rim of each jar with a warm washcloth dipped in distilled white vinegar. Set a lid and ring on each jar and hand tighten.
5. Set jars in the pressure canner, lock the pressure canner lid and bring to a boil on high heat. Let the canner vent for 10 minutes. Secure the vent and continue heating to achieve 11 PSI for a dial gauge and 10 PSI for a weighted gauge. Work quart jars for 1 hour 30 minutes and pint jars for 1 hour 15 minutes.

Nutrition:

Carbohydrates: 2 g

Fat: 13.2 g

Protein: 8.8 g

Calories: 298

12. Canned Goulash

Preparation time: 13 minutes

Cooking time: 45 minutes

Servings: 5 pint jars

Ingredients:

- 4 pounds stewing beef, cut into chunks
- 20 peppercorns

- 3 bay leaves
- 2 teaspoons caraway seeds
- 1/3 cup vegetable oil
- 3 onions, chopped
- 1 tablespoon salt
- 6 celery stalks, chopped
- 4 carrots, peeled and chopped
- 2 teaspoons mustard powder
- 1 1/2 cups water
- 1/3 cup vinegar

Directions:

1. Sterilize the bottles in a pressure canner. Allow the bottles to cool.
2. Set the meat in a bowl and add in the peppercorns, bay leaves, and caraway seeds. Massage the beef and allow marinating for an hour in the fridge.
3. Warmth oil in a saucepan over medium flame. Sauté the onions for one minute until fragrant and stir in the seasoned beef. Flavor with salt to taste before adding the rest of the ingredients.
4. Close the lid and bring to a boil for 5 minutes. Simmer for 15 minutes. Turn off the heat and allow cooling slightly.
5. Transfer the mixture to the bottles.
6. Remove the air bubbles and close the lid.
7. Place the jars in the pressure canner. Set in a pressure canner and process for 25 minutes.

Nutrition:

Calories: 627

Protein: 80.9 g

Carbs: 11.2 g

Fat: 29.2g

Sugar: 4.4g

13. **Canned Meatballs**

Preparation Time: 20 Minutes

Cooking Time: 30 Minutes

Servings: 5 half pint jars

Ingredients:

- 2 pounds ground meat
- Herbs of your choice
- 2 teaspoons salt
- Enough tomato juice to cover the meatballs

Directions:

1. Sterilize the bottles in a pressure canner. Allow the bottles to cool.
2. Set meat in a bowl and stir in the herbs and salt. Mix until well combined.
3. Boil enough water in a saucepan. Make balls out of the ground meat mixture and gently drop them into the boiling water. Allow to cook for 5 minutes then strain the meatballs.
4. Gently pack the meatballs inside the sterilized bottles. Pour in enough tomato juice over the meatballs. Leave an inch of headspace.
5. Remove the air bubbles and close the lid.
6. Place the jars in the pressure canner. Set in a pressure canner and process for 25 minutes.

Nutrition:

Calories: 272

Protein: 35.8g

Carbs: 0.8g

Fat: 14g

Sugar: 0g

14. Meat and Bean Chili

Preparation Time: 20 minutes

Cooking Time: 1 hour 40 minutes

Servings: 16-pint jars

Ingredients:

- 1 cup dried black beans (8 ounces)
- 1 cup dried kidney beans (8 ounces)
- 1/2 cup dried pinto beans (4 ounces)
- 5 pounds ground beef
- 2 pounds Italian sausage
- 1 large onion, finely sliced (2 cups)
- 1 medium green bell pepper, finely sliced (1 cup)
- 8 garlic cloves, minced
- 2 jalapeño peppers, seeded and finely sliced
- 1 cup chili powder
- 1/4 cup dried parsley
- 1/4 cup ground cumin
- 4 teaspoons coarse sea salt (optional)
- 1 teaspoon cumin seeds
- 1 to 2 teaspoons red pepper flakes
- 6 to 8 drops Tabasco sauce
- 1 teaspoon ground black pepper

- 24 medium Roma tomatoes, chopped (12 cups)

Directions:

1. The beans should be rehydrated prior to adding to chili. Sort the dried beans, detaching any damaged beans, debris, and rocks. Thoroughly wash the beans in a colander in the sink to remove any dirt. Set the dried beans in a large stockpot with enough water to cover the beans by 3 inches. Set to a boil, and then boil for 10 minutes, cover, and let sit for 1 hour.
2. In a thick-bottomed stockpot, cook the ground beef and sausage on medium-high heat for about 10 minutes or until cooked through. Set off the grease and return to a clean stockpot.
3. Attach the onion, bell pepper, garlic, jalapeños, chili powder, parsley, ground cumin, salt red pepper flakes, cumin seeds, Tabasco, and pepper to the meat mixture. Swirl well and cook on medium-heat until the onion is tender, about 5 minutes.
4. Clear out the beans in a colander and add to the meat mixture with the tomatoes. Mix well and set to a boil on medium-high heat. Set the heat and boil gently for 10 to 15 minutes, stirring often.
5. Set the hot chili into hot jars leaving a generous 1 inch of headspace. Remove any air bubbles and attach additional chili if needed to maintain the headspace.
6. Clean the rim of jar with a washcloth dipped in distilled white vinegar. Set a lid and ring on each jar.
7. Set jars in the pressure canner, lock the pressure canner lid, and bring to a boil on high heat. Let the canner vent for 10 minutes. Secure the vent and continue heating to achieve 11 PSI for a dial gauge and 10 PSI for a weighted gauge. Perform quart jars for 1 hour 30 minutes and pint jars for 1 hour 15 minutes.

Nutrition:

Calories: 377

Fat: 6g

Carbs: 6.2g

Protein 3g

15. Raw-Packed Roast Beast

Preparation time: 5 minutes

Cooking time: 1 hour and 30 minutes

Servings: 1 quart jar

- Ingredients:
- 1 pound (454 g) roast
- 1 clove garlic
- 2 small cooking onions, halved
- salt and pepper, to taste
- water, as needed

Directions:

1. Place a hunk of roast in each quart jar.
2. Attach garlic, onions, salt, and pepper to each jar.
3. Spill water into the jars over the meat and veggies. Using a rubber spatula, run it down the sides of the jars to remove any air pockets, then add more water if needed. Set 1 inch of headspace.
4. Use a cloth with vinegar on it to clear the lip of the jars. Lid the jars.
5. Using your pressure canner, work the jars for 90 minutes at 10 PSI, adjusting for altitude.

Nutrition:

Calories: 100

Carbs: 19g

Fat: 0g

Protein: 7g

BOOK 6: WATER PURIFICATION AND RESOURCE MANAGEMENT

Water is an essential element in sustaining life, acting as a vital resource for both humans and ecosystems. Access to drinkable water is crucial for our physical and mental health, as well as for maintaining the health of the ecosystems we inhabit. This chapter on water purification and filtration techniques underscores the importance of ensuring access to clean water and explores the various methods used to secure this vital resource.

Clean water is indispensable for life. It plays a critical role in numerous bodily functions and supports the growth of plant and animal species. Beyond hydration, clean water is necessary for cellular functions and the overall health of all living organisms. It is our collective duty to preserve the purity of water.

The importance of clean water is particularly apparent when considering its impact on human health. Contaminated water sources can harbor a wide range of microorganisms and pollutants, making waterborne diseases a significant threat to communities. Thus, easy access to clean water is not merely a convenience but a fundamental aspect of public health and well-being.

Moreover, the availability of clean water has implications for ecosystem resilience beyond just human-centered concerns. Many species thrive in our oceans and rivers, forming a complex web of interdependence. Pollution and contamination disrupt this balance, compromising environmental resilience and accelerating biodiversity loss. Therefore, maintaining clean water is essential not only for human health but also as a safeguard for the diverse life forms in aquatic ecosystems.

The importance of safe drinking water extends beyond individual households, influencing entire communities and having profound global implications. While everyone has the right to clean water, this right is not equally accessible everywhere. Addressing water inequality requires the implementation of strategies such as empowering community networks, managing water resources sustainably, and enforcing laws related to water purification technologies. These measures aim to ensure that all communities, regardless of location, have access to safe drinking water, thereby enhancing public health, boosting economic productivity, and promoting social stability.

Despite water covering the majority of the Earth's surface, the availability of freshwater suitable for drinking is surprisingly limited. With the pressures of rising global temperatures, expanding industrial activities, and increasing population demands, the need for prudent management of this precious resource has never been more critical. Water scarcity necessitates a shift towards sustainable practices that prioritize environmental preservation while meeting human needs. Strategies include improving water use efficiency, investing in advanced water treatment technologies, and fostering global cooperation on water conservation efforts. These approaches help in maintaining the balance between safeguarding this vital resource and fulfilling the water requirements of a growing global population.

By exploring water purification and filtration techniques, we learn not only how to provide access to clean water but also how to sustain the environmental conditions necessary for the health of our planet.

I. Basic Water Purification Techniques

Water purification techniques are essential for providing safe drinking water while preserving ecological balance. These methods ensure that water is suitable for human consumption through various effective processes.

Boiling is a simple yet highly effective method to purify water. By heating water to a boiling point, most types of pathogenic organisms, including bacteria, viruses, and protozoans, are destroyed. Boiling is recommended especially in emergency situations or in areas where there is no access to more sophisticated water purification systems. It is crucial to boil water for at least one minute at sea level, adding one minute for each additional 1,000 feet of elevation to ensure all pathogens are killed.

Distillation is a robust purification method that involves heating water to create steam, which is then cooled and condensed back into liquid form in a separate chamber. This process effectively removes impurities such as minerals, salts, heavy metals, and bacteria because these substances do not evaporate with the steam. Distillation is ideal for producing highly purified water, often used in medical and industrial applications where mineral-free water is required. Despite its high energy consumption compared to other methods, distillation provides a reliable option for accessing safe drinking water, especially in areas where other purification technologies are not feasible.

II. Mechanical Filtration

Mechanical filtration employs physical barriers to capture impurities from water. This method is utilized across a spectrum of applications, from simple household water filters to complex industrial systems. It involves materials with specified pore sizes that selectively trap particles, effectively acting as a first line of defense. Mechanical filters are particularly adept at removing larger contaminants such as sand and sediment. Their efficiency makes them an essential component in a multi-stage water purification process, where they often precede more fine-grained filtration methods.

Activated Carbon Filtration

Activated carbon filtration uses a porous form of carbon that acts like a molecular magnet, attracting and trapping various contaminants, organic compounds, and hazardous chemicals. This capacity to adsorb pollutants results in water that is clear, odorless, and aesthetically pleasing. Due to its versatility, activated carbon is widely used in residential water filters to enhance the visual and sensory quality of water. It effectively removes chlorine, volatile organic compounds, and certain pesticides, making it indispensable in providing safer and more palatable drinking water.

Reverse Osmosis (RO)

Reverse osmosis (RO) is a sophisticated and highly effective filtration method. RO systems incorporate a semi-permeable membrane that allows only water molecules to pass through while blocking larger molecules and a range of dissolved substances. This technique is extremely effective in removing heavy metals, fluoride, salts, and other dissolved minerals. The high level of purification makes reverse osmosis suitable for a wide array of applications, from home drinking water systems to industrial-scale water treatment. RO not only ensures a very high purity level but also addresses concerns about waterborne contaminants that are challenging to remove through other filtration methods.

Ceramic Filtration

Ceramic filters are made from porous ceramic material, which can remove fine particulate matter, bacteria, protozoa, and cysts from water by size exclusion, without removing beneficial minerals. These filters may also be treated with silver to enhance their antibacterial properties. Ceramic filtration is particularly useful in areas without access to chemical disinfectants and in emergency situations. They are also durable, often lasting for several years before needing replacement, making them cost-effective for continual use in household water purification.

Ion Exchange

Ion exchange is a chemical process by which unwanted dissolved ions are exchanged for other ions with a similar charge. It is commonly used to soften water by replacing calcium and magnesium ions with sodium or hydrogen ions, preventing scale buildup in plumbing and improving the effectiveness of soaps and detergents. Besides softening, ion exchange is also used to remove toxic ions such as nitrate, fluoride, and lead from drinking water. Resin beads, which are used in the ion exchange process, can be regenerated multiple times, providing an efficient long-term solution for water treatment.

Granular Media Filtration

Granular media filtration uses layers of sand, anthracite, and sometimes activated carbon to remove suspended particles and some dissolved chemicals from water. This method is effective in removing turbidity, enhancing the aesthetic quality of water, and reducing pathogen loads. It is extensively used in municipal water treatment, where large volumes of water must be processed efficiently. In addition to mechanical trapping of particles, biological processes also occur within the layers, especially in slow sand filters, where a biological layer called a "schmutzdecke" forms on the surface of the sand layer, further purifying the water through microbial action.

Electrostatic Filtration

Electrostatic filtration involves the use of charged media to attract and capture oppositely charged particles suspended in water. This method enhances particle removal by leveraging electrostatic forces, which pull colloidal and smaller particles towards charged filter surfaces, effectively trapping them. Electrostatic filtration is particularly useful for removing fine particulates that traditional mechanical filters might miss, contributing to clearer and cleaner water. It's a technique often found in combination with other filtration systems to enhance overall water purity.

Gravity-Fed Filtration

Gravity-fed filtration systems utilize the natural force of gravity to pull water through filter media, eliminating the need for electrical power. These systems typically consist of one or more reservoirs stacked vertically; water moves from the top reservoir, through the filtering medium, and collects in the lower reservoir ready for use. This method is particularly advantageous in remote areas, outdoor settings, and in situations where power supply is unreliable. Gravity-fed filters can remove bacteria, protozoa, and other contaminants, making them a popular choice for both household use and outdoor activities like camping and hiking.

III. Chemical treatment

Chemical treatment utilizes chemicals like chlorine and iodine to kill microorganisms in the water. Chlorination, one of the most common methods, involves adding chlorine or chlorine compounds to water, which is highly effective against a wide spectrum of microorganisms and leaves a residual effect that helps protect water during storage. Iodine treatment, while effective, is generally used for personal or small-scale water purification because it can leave an aftertaste and is not recommended for pregnant women and people with thyroid problems. Careful management of dosages is crucial to avoid over-chemicalization, which can lead to adverse health effects and environmental impacts.

Chlorination

Chlorination is a transformative method in water purification, essential for maintaining public health by disinfecting water supplies. This process involves adding chlorine or chlorine derivatives to water to effectively neutralize a wide range of pathogens, including bacteria, viruses, and protozoans that cause diseases like cholera, dysentery, and typhoid. Chlorine is particularly valued for its ability to maintain residual protection, meaning it continues to disinfect as water moves through distribution systems, preventing microbial regrowth. The widespread adoption of chlorination has been a cornerstone in modern water treatment, credited with significantly reducing the incidence of waterborne diseases globally. Additionally, it is economically viable, which makes it accessible for both developed and developing regions, ensuring broad implementation across various scales of water treatment facilities.

Oxidation-Reduction Reactions

Oxidation-reduction (redox) reactions are a fundamental chemical process used in water purification to remove contaminants that are difficult to filter physically. Substances like ozone, potassium permanganate, and hydrogen peroxide are introduced into water to induce chemical reactions that convert harmful pollutants into harmless or less harmful forms. Ozone is highly effective against virtually all pathogens and can degrade organic contaminants, improving both the safety and taste of water. Potassium permanganate excels in managing taste and odor problems, while hydrogen peroxide provides an environmentally friendly oxidation without leaving toxic residues. These oxidation agents are especially useful in treating industrial wastewater and in remediation efforts where traditional filtration is insufficient. The strategic use of redox reactions enhances the versatility of treatment methods, allowing for tailored solutions to specific water quality issues.

Coagulation and Flocculation

Coagulation and flocculation are critical stages in the water treatment process that work together to enhance the efficiency of sedimentation and filtration. During coagulation, small particles suspended in water become destabilized through the addition of coagulants such as aluminum sulfate or ferric chloride. This chemical addition neutralizes the charges on the particles, allowing them to come together during the flocculation stage. Flocculation follows coagulation and involves the addition of polymers that encourage the small destabilized particles to aggregate into larger clumps, or flocs. These flocs are more readily removable by sedimentation and filtration processes due to their increased size and weight. The effectiveness of coagulation and flocculation is particularly noted in the removal of turbidity, organic compounds, and pathogens from raw water, making these processes indispensable in the production of clear, safe drinking water. They are widely used in both municipal and industrial water treatment plants to ensure that water meets regulatory standards and is aesthetically pleasing to consumers.

Adsorption Process

Adsorption is a critical process in water purification that involves contaminants adhering to the surface of a material rather than being absorbed into it. This method utilizes adsorbents like activated carbon, silica gel, and others, which have large surface areas and porous structures ideal for capturing a variety of impurities. Activated carbon, for example, is especially effective at removing organic compounds, chlorine, and odors from water, making it a staple in both household and industrial water treatment systems. Silica gel excels in adsorbing substances that might cause turbidity, enhancing the clarity and quality of the treated water. The adsorption process is vital for removing pollutants that are hard to eliminate through conventional filtration, providing an additional layer of purification that enhances overall water quality.

Polyphosphate Stabilization

Polyphosphate stabilization is a chemical technique used to manage water hardness and prevent scale formation in pipes and plumbing systems. This process involves adding polyphosphates to water, which sequester calcium and magnesium ions—minerals responsible for hard water. By doing so, polyphosphates inhibit the formation of scale and deposits, which can clog pipes, reduce the efficiency of boilers and heaters, and decrease the lifespan of plumbing infrastructure. This method is particularly important in areas with hard water and is widely used in both municipal water treatment and in industrial settings to ensure smooth operation and maintenance of water delivery systems.

Silver and Copper Ion Treatment

The use of silver and copper ions in water treatment is an effective method to control microbial growth. These ions possess strong antibacterial properties, disrupting the metabolic processes of bacteria and algae, ultimately leading to their death. The process is non-toxic to humans at the concentrations used in water treatment, making it a safe option for maintaining water quality. Silver and copper ions are commonly employed in swimming pools, spas, and water storage facilities to prevent the growth of microorganisms without the need for harsh chemicals. This method not only ensures the sanitary quality of the water but also extends the life of the water systems by preventing biofilm formation and microbial-induced corrosion.

IV. UV and Solar Purification

UV and solar disinfection are innovative methods that leverage the natural power of the sun to purify water, effectively killing bacteria and other microorganisms without harming the environment. These techniques have gained attention for their ability to provide clean drinking water in a sustainable manner, using renewable resources without the need for chemicals.

Ultraviolet (UV) Disinfection

Ultraviolet disinfection employs UV light, specifically within the spectrum of 200 to 300 nanometers, to inactivate all microorganisms present in water. This range is highly effective at disrupting the DNA and RNA of pathogens, including viruses and bacteria, which prevents them from replicating and causing disease. The process involves water passing over or near UV lamps, where exposure to UV light renders the microorganisms harmless. One of the key advantages of UV disinfection is its non-chemical nature, ensuring that no harmful byproducts are produced, unlike some chemical disinfection methods. This makes UV disinfection an environmentally friendly option, widely used in both residential and industrial water treatment systems. Its effectiveness, coupled with minimal environmental impact, underscores its growing popularity in achieving high standards of water purity without additional pollutants.

Solar Disinfection (SODIS)

Solar Disinfection, or SODIS, is an innovative method that harnesses the sun's natural energy to purify water, making it an ideal technique for regions blessed with abundant sunlight. The process involves using clear polyethylene terephthalate (PET) bottles, which are filled with water and placed in direct sunlight for several hours. The mechanism behind SODIS is a combination of ultraviolet (UV) radiation and thermal treatment provided by the sun. UV radiation penetrates the water and damages the DNA of pathogens, rendering them inactive and unable to reproduce, while the heat generated enhances this effect, particularly in warmer climates.

This method has been widely embraced in many developing countries due to its simplicity and cost-effectiveness. It requires no complex infrastructure or significant financial investment, only clear PET bottles and sunlight, making it a sustainable choice for communities with limited resources. Moreover, SODIS is environmentally friendly; it uses renewable energy and minimal materials, which aligns with global efforts towards sustainable development.

While Solar Disinfection (SODIS) is an effective and accessible method for water purification, it comes with several considerations that need to be addressed for optimal use. Firstly, the effectiveness of SODIS is heavily dependent on weather conditions. It relies on clear skies and direct sunlight to function properly, and during cloudy or overcast days, the process may be less effective, potentially requiring extended exposure times or the need for alternative purification methods. Additionally, the success of SODIS depends on the availability of suitable PET bottles. In some regions, these bottles may not be readily accessible, which can pose challenges to the widespread adoption of the method. Ensuring a consistent supply of appropriate containers is crucial for the long-term viability of SODIS. Finally, public perception and acceptance are critical for the successful implementation of SODIS. Education campaigns and community engagement initiatives are necessary to dispel misconceptions about the safety and effectiveness of solar disinfection, building trust and encouraging the use of this eco-friendly purification technique. Addressing these challenges is essential for ensuring that SODIS can be effectively implemented and sustained as a reliable water purification method across various environments.

V. Collecting and Storing Water

Collecting rainwater is recognized as a long-term, ecologically friendly solution to address the problem of water scarcity. This practice involves more than just catching water; it's an efficient method that utilizes age-old wisdom to sustain human existence in harmony with the natural environment. This section explores the resurgence of traditional rainwater collection methods and their role in modern, sustainable water management.

Rainwater harvesting is based on the natural water cycle and involves collecting precipitation that falls on surfaces such as catchment areas or roofs. This process transforms a transient rainfall event into a substantial resource, offering a practical and significant way to harness natural water. The straightforward technique of rainwater harvesting not only captures rain as it falls but also builds resilience and sustainability within communities by providing a reliable water source for various needs.

Residential rainwater harvesting systems are a key aspect of this practice. These systems typically begin with guttering that channels rainwater from rooftops into downspouts, guiding it efficiently into storage containers or cisterns. Whether simple or complex, these systems transform rooftops into active water collection areas that blend seamlessly with the home's architecture. The collected rainwater, once filtered, can be used for gardening, washing vehicles, and even supplementing daily water needs, adding a level of independence to household water systems.

Components of Rainwater Harvesting System

Rainwater harvesting systems are integral for capturing and storing rainwater, providing an efficient way to manage water resources. At the heart of these systems is a well-designed setup that encompasses several key components, each playing a pivotal role in the overall functionality and efficiency of the system.

The catchment area, usually the roof of a building, is the primary surface where rainwater is collected. The effectiveness of rainwater collection largely depends on the materials used in the roof's construction and its condition. For instance, roofs made from non-absorbent materials like metal or tiles are more effective at channeling clean water compared to those covered with absorbent materials which may leach chemicals into the water. Regular maintenance of the roof is essential to ensure that the water collected is of good quality

From the catchment area, water is directed into gutters installed along the roof edges. These gutters channel the rainwater towards downspouts, which guide the water efficiently from the roof to a designated storage area. This setup is crucial for diverting water away from the building's foundation, thereby preventing potential structural damage.

Before the water enters the storage system, it passes through leaf screens and first flush diverters. Leaf screens are vital for filtering out leaves, twigs, and other solid debris, preventing them from entering the storage tanks. First flush diverters enhance water quality by discarding the initial flow of rainwater, which is likely to carry the highest load of pollutants from the catchment area. This initial diversion helps in maintaining the purity of the stored water.

The final component of the system is the storage unit, which could be tanks or barrels, designed to hold the collected water. These storage containers are available in various sizes and can be installed above or below ground. Underground cisterns are advantageous as they save space and reduce water temperature fluctuations, which helps in maintaining water quality. Above-ground barrels are simpler to install and maintain but may require more space and can be subject to temperature variations which could affect water quality.

Rainwater harvesting not only reduces dependence on municipal water supplies but also mitigates the impact of stormwater on urban drainage systems. By capturing runoff, these systems lessen the occurrence and severity of floods and reduce soil erosion. Additionally, stored rainwater can be utilized for various non-potable purposes like irrigation and washing, or it can be treated for potable uses, offering a sustainable supplement to traditional water sources.

When rainwater collection is adopted widely, it unleashes a cascade of ecological and community benefits. By using collected rainwater for outdoor tasks like gardening and other non-potable uses, households reduce their reliance on municipal water supplies, which is a significant stride toward sustainable water management. This practice can substantially decrease household water bills, as rainwater can be used for activities such as washing cars and watering plants, lessening the demand for treated municipal water for these non-essential purposes.

Additionally, rainwater harvesting systems reduce surface runoff, which in turn minimizes urban flooding risks and serves as a natural deterrent against erosion, preserving valuable topsoil from being washed away during heavy rains. The infiltration of collected rainwater back into the ground helps in replenishing aquifers, which is particularly vital in areas where groundwater is the primary source of drinking water. This aspect of rainwater harvesting aligns with broader sustainability and self-sufficiency goals, empowering individuals and communities to actively safeguard and sustainably manage their water resources.

However, rainwater harvesting comes with its challenges and considerations. Water quality concerns arise depending on the materials used in storage vessels, the size of the catchment area, and exposure to pollutants. These issues can be effectively addressed with appropriate filtration and treatment systems to ensure the safety and usability of the collected water. Regulatory environments vary, and in some places, there are stringent controls over the collection and use of rainwater. Familiarity with local regulations is crucial for integrating rainwater harvesting systems in compliance with legal frameworks.

The initial investment in a rainwater harvesting system, which includes storage tanks and filtration systems, may seem substantial, but these costs are generally offset by long-term financial and environmental benefits. Investing in these systems not only contributes to immediate savings on water bills but also enhances a community's resilience to water scarcity and supports ecological conservation efforts. Thus, while there are hurdles to consider, the numerous advantages of rainwater harvesting make it a compelling option for sustainable living.

BOOK 7: HEALTH AND WELLNESS IN A NO-GRID WORLD

I. First Aid

First aid becomes critically important in a No-Grid environment, a scenario where the absence of modern medical facilities and the failure of technology like computers and smartphones drastically change daily life. In such settings, the ability to apply basic first aid is not just useful but potentially life-saving. This shift from high-tech equipment and sterile medical environments to using simple, effective techniques and readily available materials emphasizes the crucial role of improvisation and resourcefulness in emergency healthcare.

In a world stripped of electrical power, first aid serves a dual purpose: it saves lives and fosters a sense of community and mutual support. Without the availability of quick communication methods, such as phone calls or instant messaging, individuals must develop a keen awareness of others' needs and respond effectively. Actions such as applying a comforting hand or efficiently tying a bandage take on greater significance. These not only address physical ailments but also strengthen communal bonds by building a support system far more resilient than any electrical grid could offer.

In a No-Grid situation, first aid transcends mere physical health; it becomes a symbol of human resilience and adaptability. Lacking the convenience of modern infrastructure, people must rely on their instincts and creativity to manage health emergencies. This scenario promotes quick thinking and the repurposing of everyday objects into useful tools. For example, a broken branch can be transformed into a sturdy splint, or a clean piece of cloth could serve as a compress to stop bleeding. Such adaptability is essential for survival and speaks to the profound human capacity to innovate under constraints.

Maintaining a stock of basic scientific medical supplies, such as antiseptics, sterile gauze, and pain relievers, becomes critical to manage health care effectively in energy-scarce conditions. These supplies, coupled with knowledge of traditional healing practices and natural remedies, can significantly enhance the efficacy of first aid techniques. For instance, knowing how to use honey as a natural antiseptic or creating a poultice from common herbs could be invaluable.

Recognizing our interdependence in the absence of technology is more vital than ever. Accepting the responsibility of caring for one another emphasizes the strength of human connections as a powerful healing force. This mindset is a call to action to return to the traditional wisdom that has been passed down through generations, which asserts that the strength of human connection is the greatest cure in challenging times.

Basic First Aid Skills

In the complex dynamics of daily life, it's often easy to overlook the fundamental aspects of human interaction that go beyond the noise of modern technology. Among these is basic first aid, a crucial skillset that, although sometimes overshadowed by the clamor of contemporary living, remains a profound source of competence. Unlike the cold precision of algorithms and binary code, the language of first aid is expressed through human touch and shared knowledge, bringing assurance and hope when needed most.

In an ideal world, everyone would be equipped to assist those in distress, nurturing close human connections that surpass any impersonal technological interaction. Basic first aid stands as an unsung hero in everyday life, always ready to demonstrate its value. For instance, imagine a peaceful walk in the park disrupted by a sudden scream; here, no advanced gadgetry can substitute for the profound connection and empathy required in that moment. The swift, accurate assessment and response to such emergencies are central to first aid training, emphasizing the importance of being attuned to one's surroundings and recognizing signs of distress.

Basic First Aid does not rely on sophisticated equipment but rather on simple, profound skills that anyone can learn. The act of bandaging, for example, transforms an ordinary piece of cloth into a shield against fear and harm, while the gentle touch or careful movement of the hands conveys comfort and control. Even more dramatic is cardiopulmonary resuscitation (CPR), where the rhythm of chest compressions must sync with the heartbeat, illustrating how first aid techniques weave stability back into chaotic situations.

The philosophy behind Basic First Aid extends beyond mere techniques; it is about the presence and readiness to act. It teaches us to synchronize with the life's rhythm and reminds us of our interdependence and vulnerability. The sound of a reassuring voice can guide someone through crisis better than any instruction manual, showcasing the irreplaceable value of human interaction.

Basic First Aid is fundamentally a people-centered discipline. It is about making real connections in moments of need, where a look or a gesture can communicate empathy and offer solace. Life's journey is inevitably punctuated by injuries and crises, and in these moments, it's the human touch—not flashy technology—that truly makes a difference.

Ultimately, Basic First Aid is a reflection of our shared humanity and stands as a testament to the enduring importance of interpersonal relationships and essential skills, even in our increasingly digitized world. It underscores that in times of need, the real instruments of change are not machines but people ready to lend a hand.

Basic Natural Remedies

Living in a No-Grid setting allows individuals to reconnect with nature as the constant noise of modern life fades away. This reconnection fosters a more human approach to health and wellness through the exploration of home remedies for common ailments, relying on the wisdom of nature instead of modern pharmaceuticals.

In such environments, traditional remedies take precedence. For example, honey, known for its antibacterial properties, serves as a natural treatment for cuts and scrapes. Acting as a natural antibiotic, a small dab of honey can significantly accelerate the healing process. Similarly, aloe vera, with its soothing gel, becomes a critical resource for treating burns, bites, and sensitive skin conditions. Simply slicing open a leaf releases the gel, ready to be applied directly to the affected area, showcasing aloe vera's potent healing capabilities.

For respiratory ailments like the common cold, ginger emerges as a powerful remedy in the No-Grid world. Its spicy root, when brewed into tea, can alleviate symptoms such as stuffy noses and sore throats, providing comfort reminiscent of traditional home care. The steam from ginger tea acts as a natural decongestant, proving that nature offers effective remedies that negate the need for over-the-counter drugs.

The changing seasons bring various challenges, highlighting the importance of plants like elderberry, which boosts the immune system and combats flu symptoms with its natural properties. Elderberry syrup is particularly beneficial during the colder months, reinforcing the body's defenses against seasonal ailments.

Venturing deeper into the natural world, turmeric stands out with its anti-inflammatory properties. Its curcumin content offers a natural alternative for pain relief, demonstrating how traditional knowledge can supplement or even replace modern medicines in certain scenarios. A sprinkle of turmeric not only relieves pain but also aligns with the natural cycles of the earth, emphasizing sustainability and self-sufficiency.

This journey into off-grid health and wellness is a path of self-discovery, where nature serves as a vast pharmacy and humans are its beneficiaries. The healing properties of natural remedies connect us to a time when health was intrinsic to daily life. As we explore this uncharted territory, we are reminded that beneath the complexities of modern technology, every plant holds a story of healing. By listening to the subtle sounds of nature and understanding the wisdom embedded in its core, we find that often, the solutions to our health concerns lie not in more research but in the very soil of our planet. This approach not only addresses immediate health needs but also fosters a deeper appreciation for the natural world, encouraging a lifestyle that harmonizes with the environment.

II. Maintaining Physical and Mental Health

In today's overwhelming flood of fitness media, it's time to strip back to the fundamentals and explore traditional methods of getting in shape—methods that require no exercise equipment, just you and the vast wilderness as your gym.

This guide to physical fitness challenges us to break free from the constraints of advanced technology and modern gym apparatus. It invites us to rediscover the joy of engaging our bodies in the most natural ways possible. Forget the fluorescent-lit gyms and embark on an outdoor adventure where the environment itself becomes your health coach and Mother Nature, your personal trainer. Whether running through a forest trail or walking along the beach, it's about liberating yourself from the confines of modernity.

The practice of calisthenics—using one's own body weight for limb stretching and strengthening—may seem forgotten, but it is crucial to revive it. Instead of focusing on new fitness fads, it's essential to master these basic, timeless exercises like push-ups, pull-ups, planks, and squats. These activities aren't just about building muscle; they're about returning to the fundamental movements our bodies are designed to perform

Consider how nature itself dictates physical activity. It's hard to imagine our ancestors performing isolated exercises like bicep curls. Instead, they engaged in activities that enhanced their agility, speed, and dexterity. To connect with your primal self, incorporate animal-inspired movements into your workouts, such as crab walks, frog hops, and bear crawls. This approach not only makes your exercise routine more engaging but also prevents the monotony associated with traditional gym workouts.

Mental health often teeters on the edge of neglect as we pursue success and manage the ceaseless demands of our responsibilities. It's crucial to remember that maintaining mental equilibrium is essential for overall well-being, and our thoughts play a pivotal role in this intricate balance. Practical techniques for fostering mental health include developing resilience and maintaining a calm mind, rather than relying on abstract or mystical solutions.

Prioritizing gentleness towards oneself is vital. Life is replete with challenges and mistakes, and self-compassion can alleviate the sting of disappointment more effectively than harsh self-criticism. Similarly, the act of self-reflection holds significant value—it is akin to extending empathy to a friend in distress, enhancing our emotional and mental well-being by accepting our inherent imperfections.

Mental health maintenance can be compared to enjoying a symphony where each note contributes to the harmony. Practicing mindfulness offers a refuge from a world that constantly demands our attention. This does not require elaborate rituals but rather, involves simple awareness of the present moment. Activities like savoring a cup of tea or feeling a cool breeze can help anchor us in the now, providing a peaceful escape from the noise of daily worries and concerns about the past or future.

Taking care of one's mental health requires community support, much like learning to play a musical instrument. Relationships, whether they are friendships, romantic partnerships, or family bonds, are vital for nurturing the human spirit. They help build emotional resilience through shared experiences of joy and sorrow. In a society that often values independence over connectedness, actively forming and sustaining relationships is a crucial aspect of self-care. Acts of kindness, thoughtful comments, or simply listening can significantly impact our mental health, offering comfort and reassurance in our interconnected lives.

Being creative significantly enriches our mental health. The ability to write, paint, dance, or express oneself creatively is not necessarily innate, nor is it a skill limited to a select few. Creativity involves unleashing our creative energies and recognizing that each individual has the capacity to create their own unique contributions to the world.

The role of laughter in mental health, though less frequently studied, is profoundly beneficial. A hearty laugh can alleviate feelings of despair and introduce a light-hearted perspective to life's challenges. This ability to find humor helps relieve stress and maintains an optimistic outlook

Among various strategies for mental and emotional well-being, the importance of sleep is increasingly recognized. In a society that often undervalues rest in favor of constant activity, advocating for sleep requires awareness of its benefits. Sleep is crucial as it allows the brain to recover and heal. Effective self-care includes activities that promote relaxation, such as leisurely walks, naps, or simply resting, all of which contribute to mental rejuvenation.

Mental health is a tangible aspect of life that can be improved through thoughtful choices and supportive actions. It includes cultivating self-compassion, embracing the present, fostering genuine connections, encouraging creativity, enjoying laughter, and ensuring adequate rest. Together, these elements form a comprehensive approach to sustaining mental and emotional health, creating a balanced and fulfilling life.

Finding peace can become a challenge in our hectic lives, where the demands of the digital age are relentless. In today's fast-paced world, carving out moments to sit quietly, take a deep breath, and disconnect from the noise and notifications is increasingly difficult. This is where the ancient practice of mindfulness comes into play, serving as a mental sanctuary amid the chaos.

Mindfulness, the practice of being fully present in the moment, is not a product of modern technology or a miracle cure, but rather a fundamental human capability. It involves consciously choosing to engage fully with the present moment and to be aware of our thoughts, emotions, and experiences as they occur. This ancient practice helps us navigate the maze of our thoughts and reconnect with ourselves.

Mindfulness is highly effective in reducing stress. Imagine a scenario where fear is not a constant adversary but merely a transient concern that drifts through your mind like a passing cloud. Mindfulness allows us to observe stress without becoming overwhelmed by it, offering a calm refuge in the storm of our daily lives. The key is to recognize that stress is an inherent part of life and to learn how to manage it effectively rather than trying to eliminate it altogether.

Often, we forget that our happiness is largely within our own control, especially when overwhelmed by responsibilities and demands. Mindfulness encourages us to explore the landscape of our thoughts and feelings with compassion and curiosity. Begin your journey by paying attention to the simple details of daily life, such as the sound of your footsteps, the pattern of your breath, or the flavor of food on your tongue. Through mindfulness, we can discover how to live more attuned to our experiences and cultivate a more centered, peaceful state of being.

Learning to be mindful involves slowing down to embrace the present moment, moving away from the demands of multitasking. Mindfulness teaches us to appreciate living in the now, focusing on our breathing to achieve a sense of peace and connection with our surroundings.

The appeal of mindfulness lies in its simplicity and accessibility. It doesn't require complex knowledge or elaborate systems; presence is all that is needed. This simplicity allows anyone, regardless of age, social status, or cultural background, to benefit from its practice. Mindfulness can be practiced anywhere, from the hustle of busy city streets to the tranquility of rural settings, offering a peaceful refuge from the stresses of everyday life.

While mindfulness isn't a magical solution for stress reduction, it provides valuable tools for gaining a new perspective on life's challenges. The practice aims to cultivate a resilient mind, capable of facing difficulties with composure and confidence. Although mindfulness doesn't eliminate life's problems, it equips us with the skills to handle them more effectively and lightly.

Mindfulness is more of a continuous journey than a final destination. It involves a deep exploration of our capacity for attention and presence. It is an ongoing process, where each moment of mindfulness can feel like a rejuvenating pause, allowing us to find clarity and calm within ourselves. By engaging in this practice, we can create a personal sanctuary, transcending immediate concerns and connecting with a more profound sense of peace.

III. Natural Remedies and Medicinal Plants

Foraging for medicinal herbs and plants in off-grid environments can be both rewarding and beneficial for sustaining healthcare needs. Here's a guide to help you identify, harvest, and utilize medicinal plants safely:

Identification Tips:

- Familiarize yourself with the local flora and their medicinal properties. Invest in field guides specific to your region.
- Learn to identify plants using key features such as leaves, flowers, stems, and growth habits.
- Pay attention to habitat preferences and seasonal variations in appearance.
- Be cautious of look-alike plants; misidentification can lead to harmful consequences.

Harvesting Techniques:

- Obtain permission if you're foraging on private property or protected lands.
- Harvest plants in moderation, ensuring you leave enough for natural regeneration and for other foragers.
- Use clean, sharp tools like scissors or pruning shears to minimize damage to the plant.
- Harvest parts of the plant at the appropriate time; for example, leaves are often best harvested in the morning after the dew has dried.
- Handle plants gently to prevent bruising or damaging delicate structures.
- Consider sustainable harvesting practices to preserve plant populations for future generations.

Safety Precautions:

- Verify plant identification with absolute certainty before use; consult with experienced foragers or botanists if unsure.
- Be aware of potential allergic reactions or interactions with medications.
- Avoid harvesting plants from polluted areas or those treated with pesticides.
- Always wash harvested plants thoroughly before use to remove dirt, insects, or contaminants.
- Start with small doses when using a new herb and monitor for adverse reactions.
- Know how to properly prepare and administer herbal remedies to maximize efficacy and safety.

Common Medicinal Plants for Foraging:

Echinacea (Echinacea spp.): Boosts immune function and helps fight colds and infections.

St. John's Wort (Hypericum perforatum): Used for mood disorders like depression and anxiety.

Chamomile (Matricaria chamomilla): Soothes digestive issues, promotes relaxation, and relieves anxiety.

Calendula (Calendula officinalis): Antimicrobial and anti-inflammatory; used in wound healing and skincare.

Yarrow (Achillea millefolium): Stops bleeding, reduces inflammation, and supports digestion.

Peppermint (Mentha piperita): Eases digestive discomfort, relieves headaches, and freshens breath.

Remember, while many medicinal plants offer valuable benefits, it's essential to gather knowledge and experience before foraging for them. Additionally, respect nature and its delicate balance by practicing sustainable foraging techniques and leaving minimal impact on the environment.

BOOK 8:
SECURITY AND SELF-DEFENSE STRATEGIES

I. Home Security Overview

Home security is fundamentally about protecting the sanctity of our personal spaces, blending practical measures with the emotional ties that bind us to our homes. It transcends mere physical barriers such as gates or alarm systems, touching on the less tangible aspects of feeling safe and secure in our environment. This approach to home security involves more than just installing motion detectors and cameras; it includes fostering a sense of inner peace and safety that makes a house truly a home.

Our homes are more than just structures; they are the environments where daily life unfolds, filled with family interactions and personal moments. Home security should not be seen as an afterthought but as an integral part of living spaces, contributing to the overall comfort and wellbeing of its inhabitants. It is about creating a safe haven that protects not just the physical premises but also the experiences and memories created within its walls.

The concept of home security also involves a delicate balance between openness and fortification. Security measures should be integrated seamlessly into the home environment, acting as bridges rather than barriers. This integration helps maintain the home's character and warmth, ensuring that safety mechanisms enhance, rather than detract from, the living experience.

Home security is an evolving aspect of our lives, adapting to new technologies and changing lifestyles. It includes understanding the psychological aspects of security—how it affects our feelings of safety and interacts with our emotional needs. The most effective security strategies involve a combination of technology and human presence, such as the comforting sounds of family life and the active engagement of community members.

The most fundamental protections are not always the most visible. The presence of loved ones and the laughter that echoes through the halls are just as important as any technological system in creating a secure and nurturing environment. This exploration invites us to view home security not just as a set of defensive tactics against external threats, but as a comprehensive approach to fostering a safe, peaceful, and inviting home.

Home security systems are meticulously designed to ensure robust protection and self-sufficiency during various emergencies, such as natural disasters or civil unrest. These systems integrate advanced technologies and strategies to maintain safety, sustainability, and resilience.

Key Elements of security system is robust perimeter security. This includes the installation of strong physical barriers like reinforced fencing or walls to deter unauthorized access. High-resolution surveillance cameras equipped with night vision and motion sensors play a crucial role in monitoring the perimeter continuously, ensuring that any unusual activity is detected promptly. Additionally, secure access control mechanisms, such as biometric scanners, keypads, or remote-controlled entry points, regulate who can enter the property, adding a layer of security that is both advanced and reliable.

Structural security is also paramount. Buildings are typically designed with impact-resistant materials that can withstand severe weather conditions and resist attempts at forced entry. Windows and doors are reinforced with bulletproof or shatter-resistant glass and are often made of steel or solid core materials to enhance their durability and resistance against breaches.

Energy independence is critical for survivalist systems, which often include installations of solar panels and wind turbines to ensure a continuous supply of power. These renewable energy sources are complemented by backup generators that provide emergency power during outages, running on gas, diesel, or alternative fuels, ensuring that the home remains operational even in the absence of traditional power sources.

Water and food storage are also key components of survivalist home security. Water purification systems integrated into the home filter and purify water from various sources, securing a dependable supply of clean water. Additionally, facilities designed for the long-term storage of non-perishable food items are equipped with climate control systems to maintain the freshness and nutritional value of the food stored, thereby supporting prolonged self-sufficiency.

Communication systems in survivalist homes are designed to function independently of traditional networks. This includes the use of emergency communication devices such as satellite phones and ham radios, which ensure connectivity even when conventional communication infrastructures fail. Local network systems also allow for intra-property communication, enhancing coordination and response without relying on external networks.

Lastly, self-defense capabilities are integrated into the system. Secure storage for firearms and ammunition ensures they are accessible only to authorized individuals, while also incorporating non-lethal defense options like stun devices and pepper spray. This approach not only enhances the physical security measures but also ensures that inhabitants can defend themselves if needed, while maintaining legal and ethical standards.

Together, these elements form a comprehensive and multifaceted approach to home security, designed to offer unparalleled protection and peace of mind in uncertain times.

II. Home Security Alarm Systems

Home security alarm systems are a critical component of modern residential safety, designed to detect and deter unauthorized entry into homes and alert homeowners to potential threats. These systems range from basic setups to complex networks of interconnected devices, all aimed at enhancing the safety and security of a home. Here's a comprehensive overview of how these systems work, their key components, and the benefits they provide.

The control panel acts as the central hub of a home security alarm system, coordinating all connected devices, managing the arming and disarming of the system, and facilitating communication with the monitoring service. Users can interact with the control panel through various means, such as keypads installed in convenient locations around the home, mobile devices, or even through voice-controlled smart home systems. This centralization ensures that homeowners can quickly respond to alerts and manage their security settings with ease.

Sensors are integral to the functionality of home security systems, with several types tailored to different protective needs. Door and window sensors are commonly employed to secure entry points; they trigger an alarm when these openings are disturbed while the system is armed. Motion detectors are placed in larger, open areas of the home to identify unauthorized movement, enhancing interior security. Additionally, glass break detectors add a layer of security against window breaches by detecting the specific acoustic signature of breaking glass, thereby alerting homeowners to more subtle forms of intrusion.

Surveillance cameras, both indoor and outdoor, are pivotal for real-time monitoring and recording of a home's surroundings. Modern security cameras often come equipped with features like night vision, wide-angle viewing, and motion-sensitive recording, allowing for comprehensive surveillance. Furthermore, many cameras now offer remote viewing capabilities, enabling homeowners to monitor their property from anywhere via smartphone or computer, adding a significant layer of convenience and control.

High-decibel alarms are designed to alert residents and neighbors immediately when a security breach is detected. The loud sound not only serves to notify of an intrusion but also acts as a deterrent, potentially scaring off intruders before they can cause significant harm. This feature is crucial for prompting a rapid response to secure the property.

Environmental sensors in home security systems broaden the scope of protection by detecting potential hazards like smoke, carbon monoxide, or water leaks. These sensors are essential for early warning in situations that could quickly escalate into life-threatening emergencies, providing an added layer of safety to the traditional security system.

Finally, yard signs and window decals may seem simplistic but are effective in preemptively deterring potential intruders. By clearly advertising the presence of a security system, these signs can discourage burglars from targeting the home in the first place, leveraging the power of visibility as a first line of defense.

There are two primary types of monitoring, self-monitoring and professional monitoring, each offering distinct advantages and considerations.

In a self-monitored security system, the homeowner takes on the role of monitoring the alerts generated by the system. Whenever the system detects unusual activity, such as an opened window or detected motion during an armed state, it sends an alert directly to the homeowner's mobile device, such as a smartphone or tablet. This approach allows homeowners to have direct control over the security system and decide on the appropriate response to each alert, whether it's dismissing a false alarm or contacting emergency services. Self-monitoring systems often leverage modern technology, providing apps that allow users to view live video feeds, receive instant notifications, and remotely control their home security from anywhere in the world. This type of monitoring is particularly appealing to those who prefer a hands-on approach and wish to avoid monthly fees associated with professional monitoring services.

Professional monitoring offers an added layer of security by connecting the home security system to a dedicated monitoring center that operates 24/7. When an alarm is triggered, the system sends a signal to the monitoring center, where trained security personnel assess the situation and react accordingly.

This can include contacting the homeowner to verify the nature of the alert, and if necessary, dispatching local emergency services such as police, fire departments, or medical aid. This type of monitoring is highly beneficial for providing peace of mind, particularly when the homeowner is away from home, asleep, or otherwise unable to respond to an alert themselves. Additionally, professional monitoring can potentially lead to faster emergency response times and can also qualify homeowners for discounts on homeowners insurance. However, it typically involves a monthly fee, which is a critical factor to consider when choosing this type of service.

Both monitoring options integrate with the latest home security technologies to ensure homeowners can choose the level of involvement and type of response they deem suitable for their lifestyle and safety requirements. Whether opting for the direct control of self-monitoring or the assured response of professional monitoring, the goal remains the same: to provide reliable and effective security that protects the home and its occupants.

The benefits of home security alarm systems are vast and varied, providing significant advantages to homeowners in terms of security, convenience, and financial value.

One of the primary benefits of having a home security alarm system is its ability to act as a strong deterrent to potential intruders. The visible components, such as cameras and yard signs, signal to would-be burglars that the property is well-protected, making them more likely to avoid attempting a break-in. Studies have shown that homes without security systems are significantly more likely to be targeted by criminals. The psychological impact of an alarm system can be a powerful tool in preventing crime before it happens.

Beyond mere deterrence, home security systems provide crucial alerts that help protect residents from a variety of threats. When an intrusion occurs, the system triggers an alarm that can immediately alert homeowners and neighbors, potentially stopping the intruder in their tracks. Additionally, the integration of environmental sensors adds a layer of protection, detecting potential dangers such as smoke, fire, carbon monoxide, or water leaks. This immediate notification allows for swift action to be taken, potentially saving lives and preventing property damage.

Another significant benefit is the peace of mind these systems offer to homeowners. Knowing that your home is monitored and protected around the clock, whether you're at home sleeping or away on vacation, can provide a sense of security that is invaluable. This comfort is especially important in today's fast-paced world, where concerns about personal safety and property protection can be a significant source of stress.

Installing a home security system can also increase the value of your property. Buyers often value the added security that a professionally installed and monitored system can provide. This can make a property more attractive on the market, often leading to higher resale prices. Real estate agents frequently highlight security features in listings as they are appealing attributes that can differentiate a home from others in the area.

The modern advancements in home security technology have introduced the ability to monitor and control your home remotely. This feature allows homeowners to check in on their properties via smartphones, tablets, or computers, no matter where they are in the world. You can view live camera feeds, receive alerts, and even manage system settings from afar. This level of control not only enhances security but also adds a layer of convenience that modern homeowners appreciate.

When selecting a home security alarm system, consider factors like the size of your home, the level of security needed, integration with other smart home devices, and whether DIY or professional installation suits your skills and budget. Additionally, evaluate the credibility of security providers, their service packages, and customer support services.

Home security alarm systems continue to evolve, incorporating advanced technologies like AI and machine learning to offer more precise and adaptive functionalities, ensuring that they remain an essential part of home safety strategies.

III. Self-Defense Basics

The foundations of self-defense equip us with the knowledge and skills necessary for personal protection, emphasizing the use of awareness and preparedness. Self-defense is more than just physical combat; it's about fostering resilience and ensuring one's safety through strategic thinking and environmental awareness.

Self-defense is fundamentally about controlling one's own safety. It is not merely a set of techniques for physical confrontation but a comprehensive approach that includes understanding one's instincts and emotions. The process involves debunking the myth that self-defense is solely about force or aggression. Instead, it integrates the mental and physical aspects, where understanding personal instincts and emotional cues plays a crucial role.

A key component of effective self-defense is situational awareness — not just being perpetually on high alert, but being attuned to one's environment and able to respond appropriately to potential threats. This means recognizing and adapting to the nuances of your surroundings, which can enhance your ability to protect yourself.

Mental resilience is as critical as physical capability in self-defense. Managing anxiety and fear effectively is essential, as these emotions can either hinder or enhance your ability to respond to threats. The goal is to transform fear from an adversary into an ally that informs sensible decision-making.

When discussing defense techniques, it's important to view them not just as physical tactics but as skills that require finesse and strategic thinking. These techniques should be seen as a dance between defender and attacker, where the aim is to neutralize the threat without escalating violence. This approach focuses on evasion and deflection, emphasizing the preservation of safety over confrontation.

Physical conditioning and martial arts training are vital components of a comprehensive self-defense strategy. However, the focus should be on enhancing one's overall capability rather than merely developing a formidable physical presence. True mastery in self-defense involves a harmonious balance between the body, mind, and spirit, highlighting that strength comes not only from physical power but also from mental and emotional alignment.

Effective self-defense begins long before any physical contact occurs. The foundational element is situational awareness, which involves being consistently observant and vigilant about your surroundings. This means not only scanning for suspicious behavior or individuals but also being aware of environmental factors like potential escape routes and safe zones. By maintaining an acute awareness of your environment, you can often avoid or deter potential threats before they escalate.

Alongside situational awareness, verbal assertiveness plays a crucial role in self-defense. It involves using a strong and firm voice to command an aggressor to back off or to loudly call for help. This tactic can not only deter the attacker by drawing attention to the situation but also assert your confidence and control. Verbal assertiveness is particularly effective because it helps to establish boundaries and can often prevent a situation from escalating into physical violence.

De-escalation techniques are another vital aspect of self-defense, focusing on resolving a conflict without resorting to physical confrontation. These techniques involve maintaining a calm demeanor, using non-provocative body language, and speaking in a way that aims to calm the aggressor. For example, agreeing with some points the attacker makes or offering a non-threatening compromise can diffuse the tension. The goal here is to decrease the aggressor's emotional and aggression levels, thus avoiding a potentially dangerous escalation.

In scenarios where danger escalates, prioritizing escape is often the safest course of action. This involves quickly assessing the environment for an exit route or moving towards areas with more people which can deter an attacker due to the potential of attracting attention and help. Learning to recognize the right moment to run and having the physical capability to do so efficiently is crucial. Moreover, creating distractions can provide the necessary moment to make an escape. This could involve throwing objects to divert the attacker's attention or using a loud noise to disrupt the situation and seize the opportunity to get away.

When escape is not possible, striking effectively becomes essential. Knowing the anatomical vulnerabilities of an attacker can increase the effectiveness of your defense. Targeting sensitive areas such as the eyes, nose, throat, groin, and knees can incapacitate an attacker long enough to allow for an escape. Techniques such as palm strikes, elbow strikes, and knee kicks are particularly effective due to their power and the limited training required to execute them properly. For instance, a well-aimed palm strike to the nose or a knee to the groin can cause considerable pain and reflexive withdrawal, providing a critical window to flee.

Alongside striking, being able to defend yourself from an attacker's blows can prevent injury and create opportunities to counterattack. Defensive techniques include parrying or blocking strikes, which not only protect you but also position you advantageously for a counter-strike. Learning how to break free from holds and grabs is another crucial skill. Techniques like the wrist release or using leverage against an attacker can be life-saving. Practicing these maneuvers regularly helps build muscle memory, allowing for quicker and more instinctive reactions under stress.

In self-defense, even everyday objects can become tools for protection, transforming ordinary items into means of defense in threatening situations. Items like keys, pens, or bags aren't just daily necessities but can also serve as improvised weapons. For example, keys can be held between the fingers to create a makeshift knuckle duster that increases the force of a punch, while a pen, being pointed and sturdy, can be used to strike at sensitive areas such as the eyes or throat. Bags, heavy enough or with hard objects inside, can be swung to create distance or impact against an attacker. Additionally, everyday chemicals like perfume or hairspray can be sprayed into an assailant's eyes, temporarily impairing their vision and providing a critical moment for escape. These techniques highlight the importance of adaptability and quick thinking in self-defense scenarios.

Furthermore, regular participation in self-defense classes is highly beneficial. These classes offer structured training in a variety of techniques that can include martial arts fundamentals, tactical defense methods, and scenario-based drills which prepare individuals for real-life situations. More than just physical training, these classes help build mental preparedness and situational awareness, essential components of personal safety. They provide a safe environment to practice responses to different types of confrontations, thereby increasing confidence and proficiency. Regular practice ensures that the body and mind can react swiftly and effectively when confronted with danger, making self-defense training an invaluable investment in personal security.

Martial arts have a profound history and cultural significance that extend far beyond their combat origins. They are more than just a form of exercise; they represent a dynamic expression of physicality, combining strength, agility, and the fluid harmony of movement. This discipline is not merely about adhering to a set of rigid techniques but about embracing a holistic approach to physical and mental wellness.

Martial arts foster a deep commitment to physical well-being that goes beyond the conventional gym environment, using the real world as a training ground. Practitioners view their bodies as instruments of art, with every movement contributing to their growth in skill and understanding. Unlike the finite operations of artificial intelligence, martial arts emphasize the continual development of both mind and body.

At its core, martial arts training offers foundational self-defense skills while also serving as a catalyst for personal growth and discipline. This training transcends the mechanical memorization of moves, encouraging practitioners to engage deeply with their body's signals, breathing rhythms, and internal strength.

The legal frameworks that govern self-defense are not merely rules to be memorized but are principles to be internalized and expressed through one's actions. Martial arts teach practitioners to set personal boundaries and ensure safety through strategic movement and control, utilizing an opponent's momentum to redirect energy effectively.

Martial arts and physical fitness are intrinsically linked; the training is about more than enduring repetitive exercises. It involves a dynamic and engaging approach to building strength, agility, and resilience—qualities that are crucial in everyday life as well as in specialized physical pursuits.

IV. Tools of Self Defense, Protection and Other Tools

When the grid goes down, whether due to natural disasters, societal collapse, or other emergencies, self-reliance becomes crucial. Having the right tools for self-defense, protection, and other essential survival tasks can make the difference between life and death. This guide covers the key tools you should have in your survival kit.

Self-Defense Tools

1. Firearms:
- Handguns: Compact, easy to carry, and effective for close-range defense.

- Rifles: Ideal for long-range defense and hunting. Common choices include AR-15, AK-47, and various bolt-action rifles.
- Shotguns: Versatile for both defense and hunting, particularly effective in close-quarters.

2. Ammunition:
- Stockpile a variety of ammunition types suitable for your firearms. Ensure you have enough for defense and hunting.
3. Knives:
- Fixed-Blade Knives: Durable and reliable for combat and utility purposes.
- Folding Knives: Compact and convenient for everyday carry.
- Machetes: Useful for both defense and clearing vegetation.
4. Non-Lethal Weapons:
- Pepper Spray: Effective for self-defense without lethal force.
- Tasers/Stun Guns: Provide a means to incapacitate an attacker.
5. Protection Tools
- Body Armor
- Bulletproof Vests: Varying levels of protection, from concealable soft armor to heavier tactical vests.
- Helmets: Protect against head injuries in hostile situations.
6. Home Security:
- Reinforced Doors and Windows: Make entry points harder to breach.
- Alarm Systems: Even without a grid, solar-powered or battery-operated alarms can alert you to intruders.
- Perimeter Fencing: Physical barriers to deter intruders.
7. Communication Tools:
- Two-Way Radios: Essential for communication within a group or family.
- Emergency Radios: Hand-crank or solar-powered radios to stay informed of news and weather updates.

Other Essential Tools

1. Shelter and Fire:
- Tents and Tarps: Portable shelters to protect against the elements.
- Fire Starters: Matches, lighters, ferro rods, and magnesium blocks.
2. Water and Food:
- Water Filters and Purifiers: Ensure a supply of clean drinking water.
- Canned and Freeze-Dried Foods: Long shelf-life foods for sustenance.
3. First Aid Kits:
- Comprehensive kits including bandages, antiseptics, medications, and tools for treating injuries.
4. Multi-Tools and Hand Tools:
- Multi-Tools: Versatile for various small tasks and repairs.
- Hand Tools: Axes, saws, shovels, and hammers for building and maintenance.
5. Lighting:
- Flashlights and Headlamps: Battery-operated and rechargeable options.

- Lanterns: Solar-powered or battery-operated for area lighting.
6. Navigation:
- Maps and Compasses: Reliable, non-electronic navigation tools.
- GPS Devices: Battery-operated or hand-crank powered for precise location tracking.

In a no-grid survival scenario, being prepared with the right tools for self-defense, protection, and essential survival tasks is paramount. Prioritize acquiring, maintaining, and knowing how to use these tools effectively to ensure your safety and resilience in challenging situations. Remember, preparedness is not just about having the tools but also about the skills and mindset to use them effectively.

BOOK 9:
NAVIGATING AND COMMUNICATING IN OFF-GRID SCENARIOS

I. Land Navigation Techniques

Learning to analyze maps is an essential skill that goes beyond merely interpreting symbols and lines on paper. It involves a deep understanding of geographic information, where each contour and symbol on the map translates into real-world terrain features. Analyzing a map effectively requires integrating instinct with the factual data presented, serving both practical navigation needs and broader geographical insights.

When unfolded, a map offers more than a flat depiction of land; it presents a narrative of the landscape. Contour lines on a map do not just represent elevation changes; they illustrate the varying terrains that one might encounter, such as mountains to climb or rivers to cross. Each element on the map guides the explorer, allowing for a richer understanding of the physical world.

The concept of scale is crucial in map reading. By zooming in, one can scrutinize specific paths or landmarks, while zooming out provides a broader view of extensive landscapes, effectively compressing vast terrains into manageable views. This scalability transforms how we perceive and navigate spaces, bridging large distances and diverse geographies.

Collaboration between the map user and the cartographer is fundamental. Understanding a map's contour lines, for example, helps to visualize the three-dimensional aspect of the terrain. This interaction breathes life into the static image, offering a dynamic exploration of the highs and lows of the landscape.

Using a Compass

Using a compass is an essential navigation skill, especially valuable in areas where GPS service may be unreliable. It is crucial for outdoor activities such as hiking and orienteering and serves as an important survival skill. Here's a comprehensive guide on how to effectively use a compass:

A typical compass features a magnetic needle that freely rotates to align with Earth's magnetic field, pointing towards magnetic north. The baseplate of the compass is marked with degrees in a circle from 0 to 360, representing compass bearings. Most compasses also have a rotating bezel or azimuth ring, which is used to set and follow specific bearings.

To use a compass, start by holding it flat in your hand to allow the needle to rotate freely, ensuring it is away from metal objects or electronic devices that could interfere with accuracy. The needle usually has a red end that points toward magnetic north. It's important to note that magnetic north can differ from true north, depending on your location; this difference is known as magnetic declination.

If you know the bearing you want to travel, set this on your compass by rotating the bezel so that the desired bearing number aligns with the compass's direction-of-travel arrow. Then, turn your body while holding the compass until the red end of the needle points to the number on the bezel that represents magnetic north. Ensure the direction of travel arrow on the compass baseplate is pointing straight ahead. Walk in the direction the arrow points, keeping the needle aligned with the designated mark on the bezel to maintain your bearing.

For more accurate navigation, using a compass in conjunction with a map is highly effective. Start by orienting the map. Place your compass on the map with the edge of the compass along your intended line of travel, ensuring the direction of travel arrow points towards your destination. Rotate the map and compass together until the compass needle aligns with the map's north. Your map is now oriented with actual directions.

To set a bearing from the map, identify your current location and destination. Align the compass so its edge forms a straight line between both points. Rotate the bezel until the north on the bezel lines up with the north on the map while keeping the edge of the compass aligned between the two points. The bearing to your destination is now set on your compass. Lift the compass from the map and use the bearing you've set to navigate in the real world, following the steps previously described.

Accurate compass use also involves checking for and adjusting the magnetic declination in your area if precise navigation is critical. It's beneficial to practice in a familiar area to become comfortable using a compass before relying on it in unfamiliar terrain. Periodically check your heading while traveling to ensure you haven't drifted off course.

Analyzing Terrain and Planning a Route

Analyzing terrain and planning a route is a fundamental skill for outdoor enthusiasts and professionals alike, providing the ability to safely navigate through varied landscapes. This process starts with a thorough examination of topographical maps or digital elevation models to understand the features of the area, including hills, valleys, rivers, and potential obstacles. The contour lines on these maps are especially crucial as they reveal the elevation changes and help predict the steepness of slopes, which can impact travel speed and difficulty.

When planning a route, it's important to consider the purpose of the journey—whether it's for speed, scenic value, or avoiding difficult terrain. Factors like elevation gain, water sources, and natural shelters should also be taken into account. For instance, routes that traverse higher elevations might offer expansive views but require more energy and time due to steeper ascents and potentially harsher weather conditions.

The next step involves marking waypoints—specific points of interest or checkpoints along the route. These waypoints help in navigating through the terrain and serve as rest points or decision points to adjust the route if necessary. Using GPS technology or a compass alongside physical maps ensures accuracy in following the planned route and assists in real-time location tracking.

Weather conditions play a critical role in route planning as well. Before setting out, it's essential to check the weather forecasts to avoid being caught in dangerous conditions. Adverse weather can transform the terrain, making streams impassable or increasing the risk of avalanches or landslides on steeper slopes.

Safety is paramount, so always plan alternate routes that can be used in case of emergencies or unexpected changes in the terrain or weather conditions. Inform someone about your plans and expected return time, and carry necessary survival gear and communication devices.

In sum, terrain analysis and route planning are about understanding the physical landscape, assessing risks, and preparing for various conditions. By mastering these skills, adventurers can ensure they enjoy their journeys safely and efficiently.

II. Off-Grid Communication

Off-grid communication refers to the methods and technologies used to stay connected without relying on traditional grid-based telecommunications systems, such as cellular networks or the internet. This type of communication is crucial in remote areas, during emergencies, or for those who choose to live independently from mainstream connectivity infrastructures.

The most common forms of off-grid communication include satellite phones, which operate by connecting to satellites instead of terrestrial cell sites. This allows users to make calls and send messages from virtually anywhere on the planet, provided there is a clear line of sight to the sky. Another popular option is the use of two-way radios, also known as walkie-talkies, which facilitate short to medium-range communication through radio frequencies. These devices are especially valuable in areas without any cellular coverage and are widely used in outdoor activities, disaster response scenarios, and among communities living off the grid.

Ham radios, or amateur radios, represent a more sophisticated form of off-grid communication. They require an operator license but offer a powerful means of long-range communication that can span across continents without relying on any commercial networks. Ham radio operators can use different frequencies to communicate voice, text, and data, and they often serve as critical communication links during major disasters when other systems fail.

Another innovative approach is the use of mesh networks, which involve creating a network where each device acts as a node that can transmit and receive its own signals. When one device sends a message, it hops from one node to another until it reaches its destination. This method is highly scalable and can operate independently of the internet, growing stronger as more nodes become part of the network.

For text-based communication, some turn to low-tech solutions such as the use of manual typewriters and physically transporting written messages, which, while outdated, can still be effective in certain contexts.

Moreover, technology advancements have led to the development of portable devices that can create localized networks to share messages and data directly between smartphones without needing cellular data or Wi-Fi. These devices can be particularly useful in large-scale events or remote expeditions where traditional communication infrastructures are nonexistent.

Radio Communication

Off-grid radio communication refers to the use of radio devices to transmit and receive messages without relying on traditional power grids or centralized networks. This form of communication is invaluable in remote areas, during emergencies, or in any situation where conventional communication methods are unavailable or unreliable.

At the heart of off-grid radio communication is the use of portable or stationary radio transmitters and receivers that can operate on batteries, solar power, or other independent energy sources. These radios can range from simple walkie-talkies to more complex ham radios, each serving different communication needs and distances.

One key aspect of off-grid radio communication is its independence from the internet and cellular networks, making it highly reliable in disaster scenarios where these systems might fail. For instance, ham radios, which require a license to operate, can use a variety of frequencies for long-distance communication and can even connect to the internet through gateways when traditional systems are down.

Off-grid radio systems often include features like AM/FM and shortwave bands, which can receive news and information from around the world without internet access. This is crucial for staying informed during widespread power outages or in isolated locations.

The versatility of off-grid radio communication also extends to its community and emergency response roles. Many enthusiasts and professionals use these radios as part of emergency preparedness plans, participating in networks that can quickly spread information and coordinate help in crisis situations.

Modern advancements have enhanced the capabilities of off-grid radios, incorporating digital technology to improve signal clarity and encryption for secure communication. Some systems now feature integrated GPS for navigation, digital displays for easier tuning, and pre-programmed channels for quick access to emergency frequencies.

Overall, off-grid radio communication is a robust and essential method of staying connected when traditional options are compromised. It provides a lifeline in emergencies, a tool for adventure in remote areas, and a platform for community interaction and coordination, underscoring its significance in both everyday and critical scenarios.

Tools and Methods for Emergency Signal Transmission

Emergency signal transmission is a critical aspect of survival in situations where quick rescue is essential. The ability to send clear, unmistakable signals can make the difference between life and death. Various tools and methods have been developed to ensure that individuals can communicate their location and need for assistance effectively, even from remote or challenging environments.

One of the most straightforward and widely recognized tools for emergency signaling is the whistle. Its piercing sound can travel significant distances and can be used to signal for help without requiring much physical energy. Whistles are particularly useful in dense areas like forests or mountains where visibility might be limited.

Another essential tool is the signal mirror. On sunny days, a signal mirror can reflect sunlight to create flashes visible from several miles away, even up to 10 miles under ideal conditions. This method requires some skill to aim the reflected light accurately towards a potential rescuer's location.

Flares are a more intense signaling method, used both during the day and at night. Handheld flares can burn brightly for several minutes, providing a vivid signal that can be seen from the ground and air. Smoke flares, particularly those that emit bright-colored smoke, are useful in daylight, creating a stark contrast against natural backgrounds.

For nighttime signaling, nothing beats the effectiveness of a strobe light. These devices produce a highly visible, flashing light that can be seen from great distances, making them ideal for attracting attention after dark. They are often included in marine survival kits but are just as useful on land.

In addition to these tools, there are also improvised methods that can be utilized in emergencies. Creating large symbols on the ground in open spaces using rocks, logs, or even creating trenches in the snow can help aerial search teams locate you. Fires can serve dual purposes as heat sources and emergency signals, especially when enhanced with green foliage to produce smoke.

Finally, modern technology has introduced more advanced options like personal locator beacons (PLBs) and satellite messengers. These devices, when activated, send a signal to a satellite network which then relays the location and nature of the emergency to local search and rescue services. They are invaluable in places where traditional signaling methods may not be effective due to geographical constraints.

III. Emergency Communications

Emergency communications are absolutely critical in off-grid survival scenarios. When traditional communication infrastructure such as cell towers and landlines are unavailable, perhaps due to natural disasters, remote locations, or infrastructure damage, alternative methods become essential for maintaining contact, coordinating resources, and requesting assistance. Here's why emergency communications are so important in off-grid survival:

Safety and Security: Effective communication enables individuals or groups to call for help in case of emergencies such as injuries, medical emergencies, or threats from wildlife or hostile individuals. It provides a lifeline for individuals to request assistance or guidance when faced with dangerous situations.

Coordination of Resources: In off-grid environments, resources such as food, water, shelter, and medical supplies may be limited. Communication allows for efficient coordination and distribution of resources among group members or with external aid organizations. This ensures that essential needs are met and that resources are utilized effectively.

Information Exchange: Communication facilitates the sharing of vital information such as weather updates, evacuation routes, safety protocols, and navigation instructions. Access to timely and accurate information is crucial for making informed decisions and adapting to changing circumstances.

Maintaining Contact with Loved Ones: During emergencies, individuals may become separated from their families or loved ones. Communication provides a means to reassure and stay in touch with family members, reducing anxiety and facilitating reunification efforts.

Enhancing Situational Awareness: By establishing communication networks, individuals can stay informed about the broader situation, including the status of nearby communities, ongoing rescue operations, or potential hazards. This awareness enables proactive decision-making and risk mitigation.

Coordinating Search and Rescue Operations: In the event of a missing person or search and rescue operation, effective communication is essential for coordinating search efforts, sharing relevant information, and optimizing resources to maximize the chances of a successful outcome.

Building Community Resilience: Strong communication networks foster a sense of community resilience by enabling neighbors to support each other during crises. Whether it's sharing resources, coordinating evacuation plans, or providing assistance to vulnerable members, effective communication strengthens community bonds and enhances overall resilience.

Accessing External Assistance: In severe emergencies, off-grid communities may require external assistance from emergency responders, government agencies, or humanitarian organizations. Reliable communication channels facilitate the transmission of distress signals, requests for aid, and coordination of rescue operations.

In off-grid survival scenarios where traditional communication infrastructure is unavailable, alternative communication devices become invaluable. Here are some examples of alternative communication devices that can be used:

Ham Radios (Amateur Radio):

Ham radios operate on designated amateur radio frequencies and can communicate over long distances, sometimes globally, without relying on cellular networks or internet infrastructure.

They are versatile and come in various forms, from handheld transceivers to base stations with external antennas.

Operators need to obtain an amateur radio license to legally operate these devices and must adhere to specific regulations and protocols.

CB Radios (Citizens Band Radio):

CB radios operate on specific frequencies allocated for public use and are commonly used for short-range communication between vehicles, homes, or base stations.

They are relatively simple to use and do not require a license for operation, making them accessible to a wide range of users.

CB radios are ideal for local communication within a limited range, typically up to a few miles, depending on terrain and antenna setup.

Satellite Phones:

Satellite phones use satellites orbiting the Earth to establish communication links, providing coverage in remote or off-grid areas where traditional networks are unavailable.

They offer voice calls, text messaging, and sometimes limited data services, allowing users to stay connected from virtually anywhere on the planet.

Satellite phones are essential for emergency communication in remote wilderness areas, maritime environments, or during natural disasters where terrestrial networks may be disrupted.

Personal Locator Beacons (PLBs) and Emergency Position Indicating Radio Beacons (EPIRBs):

PLBs and EPIRBs are distress signaling devices that transmit distress signals via satellite to search and rescue authorities.

They are compact, waterproof devices designed to be carried by individuals (PLBs) or installed on boats and aircraft (EPIRBs).

When activated, these devices transmit a unique distress signal along with GPS coordinates, enabling rescuers to pinpoint the user's location and initiate a response.

GMRS Radios (General Mobile Radio Service):

GMRS radios operate on specific frequencies allocated for short-distance communication and are commonly used for personal and business applications.

They offer higher power and range compared to FRS (Family Radio Service) radios, making them suitable for communication over longer distances, especially when equipped with external antennas.

GMRS radios require a license from the FCC (Federal Communications Commission) for operation.

Handheld GPS Devices with Text Messaging:

Handheld GPS devices equipped with text messaging capabilities, such as some models of Garmin in Reach or SPOT devices, allow users to send and receive text messages via satellite.

These devices provide two-way communication and also offer GPS tracking and emergency SOS features, making them valuable tools for navigation and emergency communication in remote areas.

Mesh Networking Devices:

Mesh networking devices create ad-hoc networks between multiple devices, allowing communication over short to medium distances without relying on centralized infrastructure.

They can be deployed in scenarios where traditional communication networks are unavailable or unreliable, enabling peer-to-peer messaging and data sharing among users within the network.

By incorporating these alternative communication devices into your off-grid survival kit or emergency preparedness plan, you can ensure reliable communication capabilities even in the absence of traditional infrastructure. It's essential to familiarize yourself with the operation and limitations of these devices and to have backup power sources, spare batteries, and appropriate antennas to maximize their effectiveness in challenging environments.

BOOK 10: MANAGING ECONOMIC AND SOCIAL CHANGES

I. Introduction to Bartering

Bartering is an ancient economic practice that involves the direct exchange of goods or services without the use of money as a medium of exchange. This system of trade can be traced back to the early human civilizations and remains relevant in various forms today, especially in communities where traditional currency is scarce or in situations where people prefer non-monetary transactions.

The concept of bartering is straightforward: individuals or groups exchange items or services that they possess for others that they need. The key to successful bartering lies in the mutual agreement on the relative value of the goods or services offered, which requires negotiation skills and a good understanding of the needs and offerings of the trading partner. Unlike monetary transactions, which rely on standardized prices, bartering deals are often personalized and can vary widely depending on the context and the parties involved.

Bartering offers several advantages, including the ability to obtain goods and services in the absence of money, which can be particularly useful in times of economic hardship or in closed economies. It also allows for more flexible and creative transactions that can tailor to the specific needs of the participants. Additionally, bartering can strengthen community ties by fostering direct interactions and mutual support among members.

However, bartering also has its challenges. It requires finding a trading partner whose needs match what one has to offer, which can sometimes be difficult and time-consuming. The lack of a standard measure of value can also complicate transactions, making it hard to determine fairness or equivalence, especially for more complex or high-value items. Moreover, bartering is not conducive to all types of goods or services—particularly those that are highly specialized or in situations where precise values are difficult to establish.

In modern times, bartering has evolved and can be facilitated by technology. Online bartering platforms and local exchange trading systems (LETS) provide structured environments where individuals can list goods and services for trade, specify what they are looking for in return, and engage in transactions with a broader network of users. These systems often use credits or points to help standardize exchanges, making it easier to equate diverse offerings and needs.

To become adept at bartering, start by assessing the value of your items or services. This requires a good understanding of both the market demand and the intrinsic value of what you're offering. It's important to be realistic about what your items are worth to avoid overestimating and thus prolonging negotiations or failing to make a deal.

Effective communication is crucial in bartering. Clear, honest, and respectful dialogue helps establish trust and opens the door to repeat exchanges. When initiating a barter, be upfront about what you're offering and what you expect in return. This transparency will help prevent misunderstandings and build a foundation for future transactions.

Negotiation is at the heart of bartering. The goal is not to 'win' but to reach an agreement that feels equitable to all parties involved. This often means being willing to compromise. A good negotiator knows how much they are willing to bend before the conversation starts, which helps in steering the negotiation to a satisfactory conclusion.

It's also essential to foster good relationships with those you barter with. In many cases, bartering relies on repeated interactions with the same individuals or networks. Being known as someone who is fair and reliable can open up more opportunities and make others more inclined to deal with you favorably.

Another important aspect is the legal and ethical considerations of bartering. Ensure that all trades are conducted in a manner that is legal and transparent. Some jurisdictions may have specific laws regarding bartering, especially for tax purposes, so it's vital to be informed about these rules to avoid potential legal issues.

Lastly, versatility and creativity can enhance your bartering skills. Sometimes, the direct swap of goods or services isn't straightforward. In such cases, thinking creatively about how to bundle items, involve third parties, or even stagger deliveries over time can make a barter possible and beneficial for all involved.

II. Introduction to Trading

Trading involves the exchange of goods, services, or financial instruments between parties, either within an economy or across international borders. At its core, trading is driven by the principle of comparative advantage, whereby parties exchange what they can produce at lower costs for what others can provide more efficiently. This process is fundamental to the functioning of modern economies and plays a crucial role in allocating resources, distributing goods, promoting competition, and fostering economic interdependence among nations.

In financial markets, trading refers to the buying and selling of securities, such as stocks, bonds, and derivatives, on exchanges or over-the-counter. This type of trading is aimed at generating profits from price fluctuations in the market. Traders may operate on various time frames, from very short-term, such as day trading, where positions are held and liquidated within the same trading day, to long-term investments spanning years.

Technological advancements have greatly influenced trading practices, introducing automated trading systems and algorithms that can execute trades at speeds and volumes unattainable by human traders. These technologies also help in managing risks and identifying opportunities through predictive analytics and real-time data processing.

Moreover, trading is not just about profit but also about risk management. Effective traders are proficient at assessing market conditions, using various tools and indicators to analyze potential moves and mitigate losses. Strategies such as diversification, hedging, and using stop-loss orders are employed to manage and spread risk.

Ethical trading practices are also crucial, ensuring transparency and fairness in markets. Regulatory bodies worldwide enforce rules and standards to prevent fraud, insider trading, and other malpractices, thereby protecting the interests of all market participants.

Ultimately, trading is a dynamic and integral part of the global economy that reflects the collective behaviors and decisions of individuals, companies, and nations. It is a complex field that blends economic principles, psychology, technology, and governance, continually evolving with changes in market conditions, regulatory landscapes, and technological advancements.

Mastering the art of trading in the financial markets requires a combination of knowledge, strategy, and psychological fortitude. Success in trading is not merely about making profitable trades, but also about consistently applying a disciplined approach, managing risks effectively, and learning continuously to adapt to ever-changing market conditions.

The first step towards mastery is building a solid foundation of knowledge. This includes understanding the basics of the markets you intend to trade, whether they are stocks, bonds, commodities, or currencies. Familiarize yourself with different trading instruments and the factors that influence their prices. This education should also cover the technical aspects of trading, such as reading charts, understanding indicators, and interpreting market data.

Developing a trading strategy is crucial. Your strategy should align with your investment goals, risk tolerance, and time horizon. It could be based on technical analysis, fundamental analysis, or a combination of both. Technical traders focus on chart patterns and statistical indicators to make trading decisions, whereas fundamental traders analyze economic indicators, company earnings, news, and other qualitative and quantitative factors. It's important to backtest your strategy with historical data to ensure its viability before implementing it in live trading.

Risk management is a critical component of successful trading. It involves setting limits on how much capital you are willing to risk on a single trade and overall. Utilizing stop-loss orders and managing position sizes can help minimize potential losses and protect your trading capital. Successful traders are those who can preserve their capital and stay in the game long enough to benefit from their trading opportunities.

Emotional discipline is equally important. The psychological aspect of trading can often be the most challenging. Traders must learn to keep emotions like fear and greed in check to make objective decisions based on their trading plan. This requires maintaining a cool head during market volatility and being patient for the right trading opportunities.

Continuous learning and adaptation are essential. The financial markets are dynamic, with new patterns and trends emerging all the time. Keeping abreast of market news, understanding global economic trends, and continually refining your trading strategy based on market feedback are key to staying relevant and successful in trading.

Networking with other traders and participating in trading communities can also provide insights and foster learning through shared experiences. Many traders find value in mentors or trading coaches who can offer guidance, improve their strategies, and help them avoid common pitfalls.

BOOK 11:
EMERGENCY PREPAREDNESS CHECKLIST

Emergencies are unpredictable, and facing a natural disaster, power outage, or any unforeseen event can be daunting. Having a well-thought-out Emergency Preparedness Checklist is crucial, as it serves as a practical tool to help you manage such situations effectively. This checklist is not merely a list of items but a vital component of your readiness strategy when unexpected events occur.

Effective emergency preparedness involves more than just accumulating supplies; it requires a comprehensive understanding of readiness. Think of it as a roadmap—a guide that integrates into your daily life, ensuring you are mentally, emotionally, and practically prepared for any challenges that may arise.

A key aspect of preparing for emergencies is adopting the right mindset. It's about embracing a philosophy of self-sufficiency, resilience, and proactive behavior. The checklist is more than a collection of tangible items; it is an essential part of your journey to developing a robust preparedness mindset.

It's important to customize your Emergency Preparedness Checklist to your specific needs. A one-size-fits-all approach is not effective when it comes to emergency preparedness. Your checklist should reflect your personal circumstances, including your location, family composition, and any special requirements you may have. This customization makes the checklist a tailored tool that guides your preparedness efforts, ensuring you are equipped to handle the unique challenges you might face.

Creating Your Comprehensive Emergency Preparedness Checklist

Your home serves as a sanctuary, especially during uncertain times. It is essential to keep your home safe and secure by storing sufficient water, non-perishable food, vital tools, and ensuring a reliable source of light. These elements are crucial for maintaining safety and comfort when it matters most.

The importance of a well-packed emergency bag cannot be overstated. It should include clothing appropriate for any weather, personal hygiene items, and important documents, making sure you're ready to evacuate at a moment's notice. This preparation not only keeps you prepared but also comfortable, regardless of the circumstances.

In today's interconnected world, having access to information is crucial, especially in emergencies. This includes essential communication tools like smartphones and radios, as well as often overlooked but vital navigation aids such as maps, compasses, and GPS devices. Being equipped with the right tools to stay informed and navigate is as critical as having basic survival items.

Beyond basic first aid, assembling medical supplies tailored to your family's specific health needs, including prescription medications and personal health devices, is vital. Mental and emotional well-being should also be considered, incorporating stress-relief aids and comfort items. This comprehensive approach ensures not only physical health but also mental and emotional resilience during emergencies.

Emergency planning must also extend to all family members, including pets, the elderly, and infants, each of whom has unique needs that must be addressed to ensure their comfort and safety. Whether they have fur or not, the well-being of every family member is paramount in crisis situations.

Bringing Your Checklist to Life

Your emergency preparedness checklist is not a static document; it is dynamic and designed to evolve. We guide you through the process of regular reassessments, ensuring that your preparedness strategy adapts alongside changes in your life. Being prepared is an ongoing commitment rather than a one-off task. This approach helps ensure that you remain ready and resilient as new risks emerge and personal circumstances shift.

This checklist is not meant to be a rigid set of instructions. We encourage you to personalize it by including items that reflect your individual needs and concerns. Whether it's a sentimental object, a handwritten note, or a detailed family emergency plan, these personal touches transform the checklist from a mere survival tool into a comprehensive plan that supports thriving even during challenging times.

As you develop your Emergency Preparedness Checklist, it's important to view it as more than just a collection of items. It represents your commitment to preparedness and acts as a tangible expression of your readiness to handle the unexpected with resilience and resourcefulness. By going beyond the basics, we aim to create a checklist that not only prepares you for emergencies but also empowers you to confidently navigate through life's uncertainties. This enhanced focus not only boosts your immediate readiness but also bolsters your long-term resilience, making you better equipped to face whatever challenges may come your way.

Here's a more detailed and organized emergency preparedness checklist specifically for situations where the power grid is down. This will ensure you have all necessary supplies and plans in place to handle extended periods without electricity effectively.

Basic Needs

Water:

Storage:

- Store at least one gallon of water per person per day for at least three days.
- Consider larger storage containers for extended emergencies.

Purification:

- Water purification tablets.
- Portable water filters.
- Methods for boiling water.

Food:

Non-perishable Items:

- Canned meats, fruits, and vegetables.
- Protein or granola bars.
- Dry cereal or granola.
- Peanut butter and jelly.
- Nuts and dried fruit.

- Crackers.
- Powdered milk.
- Comfort foods (cookies, candy).

Specialty Items:

- Infant formula.
- Pet food.

Utensils:

- Manual can opener.
- Disposable plates, cups, and utensils.

Shelter:

- Tents and tarps.
- Emergency blankets and sleeping bags.
- Warm clothing and extra blankets.

First Aid and Medical Supplies

- First Aid Kit
- Bandages of various sizes.
- Sterile gauze pads and adhesive tape.
- Antiseptic wipes and hydrogen peroxide.
- Tweezers, scissors, and safety pins.
- Pain relievers (aspirin, ibuprofen).
- Antihistamines.
- Anti-diarrhea medication.
- Prescription medications (7-day supply).
- Medical gloves and masks.

Personal Hygiene:

- Hand sanitizer.
- Moist towelettes.
- Garbage bags and plastic ties for personal sanitation.
- Toilet paper and tissues.
- Feminine hygiene products.

Tools and Equipment:

- Lighting and Power
- Flashlights and extra batteries.
- Solar-powered or hand-crank lanterns.
- Battery-powered or hand-crank radio.
- Solar chargers for electronics.

Cooking and Heating:

- Portable stove or grill with fuel (propane, butane).
- Waterproof matches or lighters.
- Fire extinguisher.
- Extra blankets and warm clothing for cold weather.

General Tools:

- Multi-tool or Swiss Army knife.
- Duct tape.
- Wrench or pliers for turning off utilities.
- Rope or paracord.
- Basic tool kit (screwdrivers, hammer, nails).
- Communication and Navigation

Communication:

- Battery-powered or hand-crank emergency radio.
- Extra cell phone battery packs or solar chargers.
- Walkie-talkies for short-distance communication.
- Whistle to signal for help.

Navigation:

- Local maps and compass.
- GPS devices with extra batteries or solar chargers.
- Important Documents and Cash

Documents:

- Copies of insurance policies, identification, and bank account records.
- Emergency contact information.
- Personal identification (passport, driver's license).

Cash:

- Cash in small bills (consider local currencies if applicable).
- Coins for payphones or vending machines.

Personal and Family Safety:

- Clothing and Bedding
- Sturdy shoes and work gloves.
- Rain gear and extra socks.
- Thermal underwear and layers for warmth.
- Extra blankets or sleeping bags.

Safety Gear:

- Masks to filter contaminated air.
- Plastic sheeting and duct tape to seal windows and doors.

Entertainment:

- Books, games, puzzles, and other activities for children.
- Deck of cards.

Pets:

- Pet food, water, medications, and a carrier or leash.
- Copy of pet medical records.

Planning and Preparation:

- Family Emergency Plan
- Designate meeting places: one near your home and one outside your neighborhood.
- Establish an emergency contact person outside your immediate area.
- Practice emergency drills regularly with your family.

Maintenance and Review:

- Regular Checks
- Rotate food and water supplies every six months.
- Test and charge equipment regularly.
- Check expiration dates on all medications and replace as needed.
- Review and update emergency plans and contacts every six months.
- Conduct regular inventory of all supplies to ensure nothing is missing or expired.

Emergency Response Items:

- Evacuation Bag
- Pack a "Go Bag" for each family member with essentials:
- Basic hygiene supplies.
- Change of clothes.
- Copies of important documents.
- Small first aid kit.
- Snacks and water.
- Flashlight and batteries.
- Emergency cash.
- Multi-tool.

By following this comprehensive checklist, you can ensure that you and your family are well-prepared to handle extended periods without power and other essential services. Regularly review and update your supplies and plans to stay ready for any emergency situation.

BOOK 12:
NAVIGATING THE UNCHARTED TERRITORIES OF OFF-GRID LIVING

As we conclude our exploration of off-grid living, we find ourselves at the nexus of self-sufficiency and the unknown. Our journey has been more than a collection of survival tactics; it has been a deep dive into a lifestyle that defies conventional norms and challenges us to reconsider what it means to be truly connected to our surroundings.

Looking ahead, the future of off-grid living presents a complex tapestry of possibilities rather than a clear-cut scenario. The essence of off-grid life—characterized by adaptability and resourcefulness—will become increasingly essential in a world facing constant change and uncertainty.

Once viewed as a niche lifestyle, off-grid living is now at the forefront of sustainable practices. This movement towards using renewable energy, minimizing environmental impacts, and emphasizing holistic well-being is not just a trend but a significant shift. This paradigm challenges us to rethink our coexistence with the planet. As we confront climate change and the depletion of natural resources, the principles of off-grid living could offer vital solutions for a more sustainable and harmonious future.

Technology is playing a paradoxical yet pivotal role in advancing off-grid lifestyles. Innovations in energy systems, sustainable agriculture, and communication technologies are bridging the gap between self-sufficiency and modern conveniences. These advancements are enhancing the off-grid experience, making it more feasible and efficient for a growing number of people who choose to live disconnected from traditional infrastructure yet remain interconnected with global developments.

The journey into off-grid living is not a static endeavor; it's a dynamic, continually evolving relationship with our environment and ourselves. As we conclude this exploration, it becomes clear that ongoing education is crucial for success in an off-grid lifestyle.

Education here goes beyond traditional classrooms; it involves continuous learning from the land, the community, and the experiences that arise. From honing survival skills to adopting new sustainable technologies, those living off-grid must be lifelong learners, always adapting and expanding their knowledge and abilities.

Modifications are also a constant element in off-grid living. The essence of this lifestyle is adaptation—adjusting to seasonal changes, unforeseen challenges, and the evolving needs of a self-sufficient home. This includes not only altering physical structures but also updating mental frameworks, as resilience and adaptability become foundational to a successful off-grid existence.

As this book closes, we step into a reality where the off-grid lifestyle is not just a destination but an ongoing journey. The future holds the promise of a world where living off the grid is not only a viable option but also a mainstream choice. This lifestyle's principles are poised to become integral to our harmonious coexistence with the planet.

Whether you are a seasoned off-grid enthusiast or someone just beginning to explore the possibilities, remember that this journey is not about reaching a final destination; it's about embracing the continuous evolution of life in sync with nature. May your off-grid journey be rich with discoveries, and may the lessons shared here help you build a sustainable and fulfilling future.

As we conclude our examination of off-grid living infrastructure, it is crucial to highlight two foundational pillars that sustain a resilient and self-sufficient lifestyle: continuous learning and adaptation.

In the dynamic realm of off-grid living, the significance of embracing new knowledge is paramount. The journey of off-grid living does not conclude with the final words of this book; instead, it continually evolves as we engage with and apply the latest techniques, technologies, and innovations that enhance our independence from the traditional power grid.

Adaptation is central to sustainability. As we navigate the unknowns of off-grid life, flexibility becomes our most valuable tool. It is essential to be willing to refine and adjust our strategies based on feedback from our community and personal experiences. The ability to adapt is not just crucial for survival; it is what allows us to thrive amidst uncertainty.

Looking forward, the future of off-grid living is filled with promise and potential. Developments in renewable energy, sustainable agriculture, and non-electric technologies are not merely enhancing this lifestyle but are proving to be necessary for a sustainable future. By adopting the principles outlined in this guide, you are not only preparing for a life off the grid; you are positioning yourself at the forefront of a movement towards a more sustainable, integrated world.

In closing, let this guide serve as your companion—a point of reference, a source of inspiration, and a reminder that the path to off-grid living is an active, enriching pursuit. May your days be enriched by the fulfillment that comes from self-sufficiency, and may your ongoing commitment to learning and personal growth illuminate the path to a brighter, more resilient future.

BOOK 13: PRINTABLE SURVIVAL GUIDES FOR SPECIFIC SCENARIOS

In a world where uncertainty prevails, the demand for comprehensive survival guides tailored to specific situations is more crucial than ever. "BOOK 13: Printable Survival Guides for Specific Scenarios" serves as an essential resource during times of disaster. This collection is not merely informational; it acts as a vital support system in crisis situations, providing knowledge designed for practical application under stress.

The value of having printable survival guides in an increasingly digital-dependent society cannot be overstated. Despite seeming outdated, the physical nature of these guides offers indispensable benefits. In scenarios where digital access may be compromised and electricity is scarce, possessing a physical guide ensures that crucial information remains accessible, providing a significant advantage in emergency situations.

This bonus installment of survival guides is meticulously assembled to address the unique challenges posed by various emergencies such as earthquakes, blackouts, and economic downturns. Each guide is not a generic handbook but a customized blueprint designed to navigate the complex details of these distinct emergencies effectively. By focusing on the specificities of each scenario, these guides equip individuals with the tools and knowledge necessary to manage and adapt to crises with confidence.

A Wealth of Downloadable Wisdom

Imagine having access to a comprehensive repository of information, readily downloadable with just a click. This resource not only provides information but also empowers users to handle various crises. Whether it's the seismic disturbances of an earthquake, the sudden darkness of a blackout, or the financial repercussions of an economic downturn, these guides serve as reliable allies in times of uncertainty.

Earthquakes: Navigating the Tremors

During an earthquake, the unsettling shaking of the ground can induce panic. However, with the earthquake survival guide from this resource, you will learn not just how to survive, but how to effectively manage in the aftermath. The guide covers everything from stabilizing your immediate surroundings to addressing first aid needs, offering a comprehensive plan to deal with seismic events.

Blackouts: Illuminating the Darkness

In the event of a blackout, when the constant presence of electricity ceases, mere candles and flashlights are not sufficient. This guide provides practical advice on navigating through the darkness. It includes information on utilizing alternative energy sources, preserving food safely without power, and maintaining communication with the outside world during power outages.

Economic Downturns: Navigating Financial Storms

Facing an economic downturn involves more than merely tightening one's financial belt. It requires a strategic approach to navigate through the financial challenges effectively. This guide acts as a financial navigator, offering strategies for protecting your assets and identifying alternative sources of income. It provides a thorough roadmap for fostering financial resilience during economic uncertainties.

Emergency Food Preservation Guide:

In situations where fresh food may not be readily available, this guide provides methods for preserving food for long-term storage. It includes techniques like canning, dehydrating, smoking, and pickling to extend the shelf life of perishable foods and some recipes for your meal planning.

Urgent Foraging and Wild Edibles Guide:

This guide provides information on identifying edible plants and mushrooms in the wild. It includes detailed descriptions, photographs, and safety tips to help you forage for food safely and effectively.

Emergency Communications:

Evaluate the nature of the emergency and determine your communication needs. Consider factors such as the extent of infrastructure damage, the availability of power sources, and the range of communication devices.

Emergency First Aid with Medicinal Plants:

Identifying medicinal herbs that can be foraged for sustenance and healthcare in off-grid environments, including identification tips, harvesting techniques, and safety precautions.

Printable survival guides are comprehensive resources designed to provide crucial information and actionable steps for effectively managing emergencies and adverse situations. These guides serve as detailed manuals that equip you with the knowledge and skills necessary to not only survive but also to adapt and overcome challenges. As you engage with these downloadable handbooks, you're adopting a mindset focused on preparedness and self-reliance, crucial for thriving in unpredictable environments.

In a world where uncertainty prevails, these printable survival publications are essential tools. They offer practical advice and techniques for navigating the complexities of survival scenarios. Each guide is crafted to help you understand and implement survival strategies that harness the resilience and adaptability of the human spirit. These resources are vital in helping you prepare for and respond to emergencies, ensuring you have the capability to handle whatever challenges you may encounter.

CONCLUSION

This book serves as an extensive guide for those seeking self-reliance and preparedness in the event of a grid-down scenario. It covers a wide range of survival strategies, emphasizing the importance of self-sufficiency through practical projects and skills. The book underscores the necessity of being prepared for various emergencies by providing detailed instructions on building, growing, and maintaining essential resources without relying on modern infrastructure. In essence, it equips readers with the knowledge and tools needed to thrive independently and sustainably, ensuring safety and resilience in times of crisis.

Made in the USA
Las Vegas, NV
21 September 2024

95606382R00077